P9-CDJ-593

PANDORA'S DNA

PANDORA'S DNA

Tracing the Breast Cancer Genes Through History, Science, and One Family Tree

LIZZIE STARK

First edition
Published by Chicago Review Press, Incorporated
814 North Franklin Street
Chicago, Illinois 60610
ISBN 978-1-61374-860-2

Library of Congress Cataloging-in-Publication Data
Stark, Lizzie.
Pandora's DNA : tracing the breast cancer genes through history, science, and one family tree / Lizzie Stark.
pages cm
Summary: "Would you cut out your healthy breasts and ovaries if you thought it might save your life? That's not a theoretical question for journalist Lizzie Stark's relatives, who grapple with the horrific legacy of cancer built into the family DNA. It is a BRCA mutation that has robbed most of her female relatives of breasts, ovaries, peace of mind, or life itself. In Pandora's DNA, Stark uses her family's experience to frame a larger story about the so-called breast cancer genes, exploring the morass of legal quandaries, scientific developments, medical breakthroughs, and ethical concerns that surround the BRCA mutations. She tells of the troubling history of prophylactic surgery and the storied origins of the boob job and relates the landmark lawsuit against Myriad Genetics, which held patents on the BRCA genes every human carries in their body until the Supreme Court overturned them in 2013. Although a genetic test for cancer risk may sound like the height of scientific development, the treatment remains crude and barbaric. Through her own experience, Stark shows what it's like to live in a brave new world where gazing into a crystal ball of genetics has many unintended consequences"— Provided by publisher.
Includes bibliographical references and index.
ISBN 978-1-61374-860-2 (hardback)
1. Stark, Lizzie—Health. 2. Mastectomy—Patients—United States—Biography. 3. Breast—Cancer—Genetic aspects. 4. BRCA genes. I. Title.

RD667.5.S73 2014
616.99'449042—dc23

2014018310

Interior design: PerfecType, Nashville, TN

Printed in the United States of America
5 4 3 2 1

For the women in my family
and the men who love them.

Contents

"All happy families are alike;
each unhappy family is unhappy in its own way."

—Leo Tolstoy

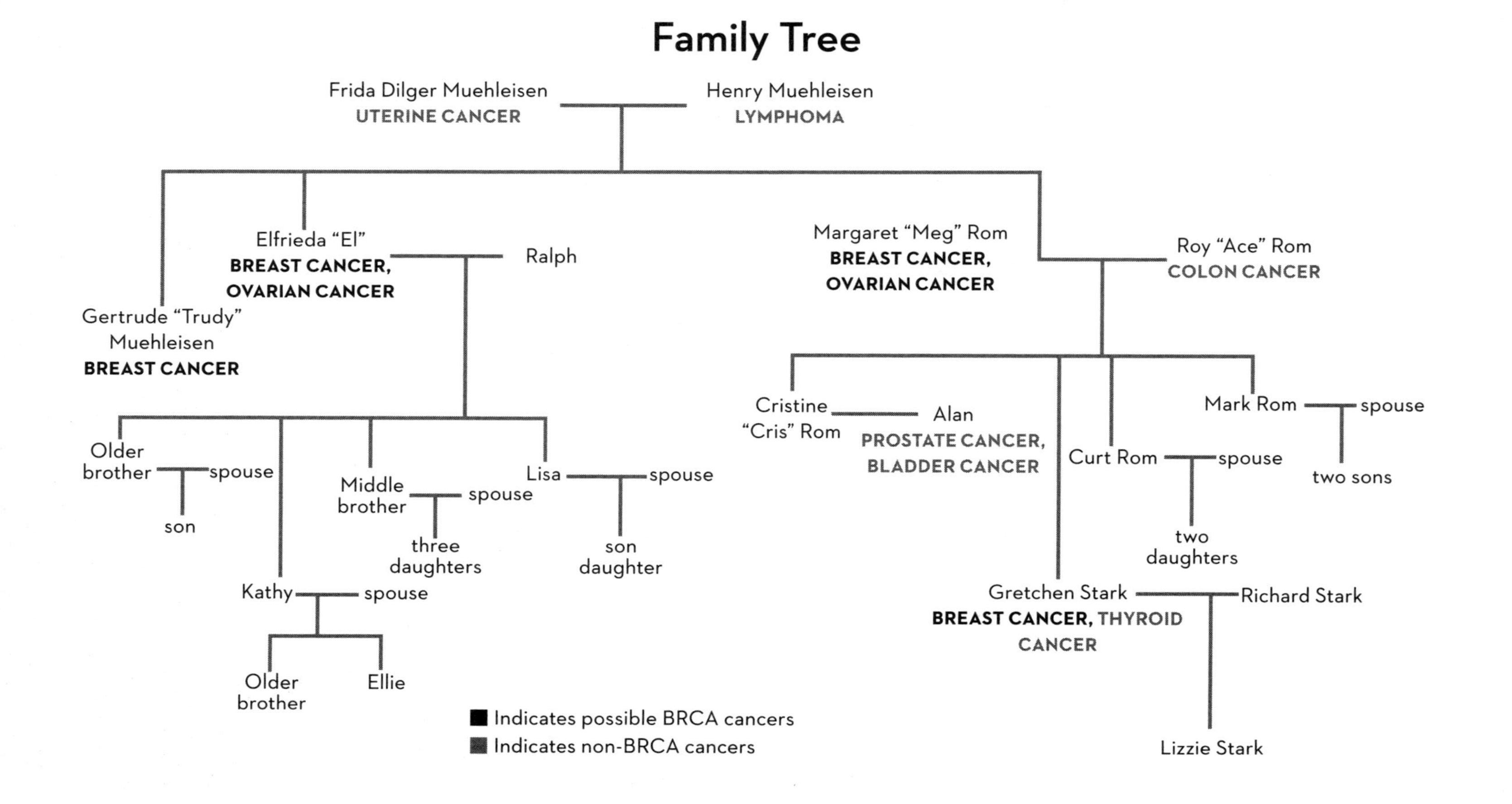
Family Tree
Frida Dilger Muehleisen
UTERINE CANCER
Henry Muehleisen
LYMPHOMA
Elfrieda "El"
BREAST CANCER, OVARIAN CANCER
Ralph
Gertrude "Trudy" Muehleisen
BREAST CANCER
Margaret "Meg" Rom
BREAST CANCER, OVARIAN CANCER
Roy "Ace" Rom
COLON CANCER
Older brother
spouse
son
Kathy
spouse
Older brother
Ellie
Middle brother
spouse
three daughters
Lisa
spouse
son
daughter
Cristine "Cris" Rom
Alan
PROSTATE CANCER, BLADDER CANCER
Curt Rom
spouse
two daughters
Mark Rom
spouse
two sons
Gretchen Stark
BREAST CANCER, THYROID CANCER
Richard Stark
Lizzie Stark
Indicates possible BRCA cancers
Indicates non-BRCA cancers

Prologue

My mother holds me on a sunny afternoon in Mesa Verde, Colorado, where we are vacationing with relatives. In the photo my father has snapped, I am an infant. My mother, dressed in a paisley muumuu and a turban, manages to smile at the camera, despite the fact that her clothing conceals a missing breast, the opening salvo in her strike against cancer diagnosed at age thirty. I do not yet know that she wears the turban because she has undertaken this two-week family vacation—a camping trip with an eighteen-month-old that seems insane to me in retrospect—between chemotherapy treatments that have taken her hair but not her fierce desire to live and see me through my childhood. I do not yet know that there has been so much cancer in her family that her sister and cousins will choose to remove healthy organs rather than risk a potentially fatal illness.

Twenty-four years later, I will carry this frightening experience—a mother sick first from breast cancer and later from thyroid cancer, the invincible god of my childhood weakened and injured by disease—to the genetic counselor with me. The waiting room walls are beige and smeared with stains from oily hair and anxious sweat. I find it terrifying that the genetic counseling department at this hospital shares an office with oncology, which is a place I desperately want to avoid. I am the youngest patient in the room by at least twenty years, except for the shadows I have dragged in here with me. The ghost of my great-aunt Trudy, dead of breast cancer at thirty-one, sits by a plastic plant and reminds me that wishful thinking is no panacea. My great-aunt Elfrieda's shade sits further off in her wheelchair, the one that invasive ovarian cancer confined her to after the disease spread to her spine. The absence of my grandmother's spirit is notable. My mother has not told her about this meeting; we don't want to pile another cancer-related ordeal on a fading elderly woman who has already lost so much—breasts, ovaries, parents, sisters, and peace of mind—to the disease. She and her two sisters collectively endured six bouts of "women's illness," as cancer of the female organs used to be called. Of the four women in my mother's generation, only one endured a breast cancer diagnosis, but all four chose double mastectomies. My story feels written before it's even begun.

The genetic counselor has to fetch extra paper in order to record all the cancer in my family history. She explains the statistics to me in a gentle voice, as if trying to tell me my puppy has just died. I don't understand why she sounds like that. "Fourteen percent chance of developing breast cancer at a minimum, if I have the gene, right?" I say. "As in one-four."

"No," she says, because I have misheard her. "As in four-zero to six-zero. Forty to sixty percent." The numbers feel like a slap in the face, shocking and unacceptable. It is the first time I've confronted these statistics, and the first time that they make me feel nauseous with dread. Other doctors will give me other percentages, explaining that my lifetime risk of breast cancer could be as high as 87 percent.

Cancer runs on my mother's side of our family because we happen to have a certain mutation on our BRCA1 gene. Every person has two pairs of BRCA genes in each cell of the body. BRCA1 is located on chromosome 17, and BRCA2 is located on chromosome 13. Scientists identified these genes in the 1990s and named them BRCA, officially for "breast cancer," but unofficially for "Berkeley, California," where the location of the first one was narrowed down. The person who named BRCA1, Mary-Claire King, had also hoped to name it for Victorian scientist Paul Broca, who recorded an early cancer family, but she was only permitted four letters. Certain types of hereditary mutations in these genes can drastically elevate a woman's chance of developing breast and ovarian cancer and certain other cancers, and can elevate a man's chance of developing male breast cancer, prostate cancer, and other cancers. Though in my family my grandmother passed down this mutation, men can also inherit a mutation and pass it to their children. A person of either sex with a cancer-linked BRCA mutation has a 50 percent chance of passing that gene along to their kids.

We don't know exactly how much a harmful BRCA mutation ups the risk of breast and ovarian cancer in women—estimates vary widely, according to Dr. Mary Daly, chair of the Department of Clinical Genetics at Fox Chase Cancer Center in Philadelphia, because they come from relatively small studies and because we are in the early days of medical genetics. The original estimates were higher, she says, because they came from early studies of the mutations: "To be in those studies you had to have at least four women [in the family] with breast or ovarian cancer," which meant the studies looked at the grimmest, most intense mutations. Hereditary BRCA1 and 2 mutations take many different forms, and each mutation comes with differing levels of cancer risk; some mutations are completely harmless, while others raise risk modestly, or spike it dramatically. "No one has done the study that you really need, which is to do a random sampling of people in America," Dr. Daly says. "So we have to rely on these smaller studies, where each of these studies has different limitations." We have to do the best we can with imperfect science.

Now, seven years after my first visit to genetic counseling, and with the science still coming in, the estimates have been revised, mostly downward. But they are still large compared to ordinary women.

	Lifetime chance of developing breast cancer	**Lifetime chance of developing ovarian cancer**
Average woman	12%	1.4%
Woman with a BRCA1 mutation	55–65%	39%
Woman with a BRCA2 mutation	45%	11–17%

The size of these numbers is important because statistics like this propel BRCA carriers* to action, and they have only three options: surveillance to help catch cancer early, chemoprevention using estrogen blockers to head cancer off at the pass and temporarily reduce risk, or surgery to remove the offending organs to permanently lower risk.

In addition to having an elevated lifetime risk of developing cancer, BRCA carriers also tend to develop cancer at younger ages compared to the general population. The median age of breast cancer diagnosis is sixty-one. In my family—admittedly a tiny sample

* Though every human being is technically a "BRCA carrier," since we all have BRCA genes in nearly every cell of our bodies, for ease of reference, in this book I will use "BRCA carrier" to mean someone who has inherited a faulty copy of the BRCA1 or BRCA2 genes.

size—it's more like thirty-four. Even cancer in one's twenties is not unheard of among BRCA families. BRCA1 breast cancers are also more likely than other breast cancers to be triple negative. Triple-negative cancers are resistant to certain therapies and "generally have poorer prognosis than other breast cancers," according to the National Cancer Institute. BRCA breast cancer is also more likely than typical breast cancer to recur in the unaffected breast.

BRCA mutations are pretty rare—about one in four hundred to one in six hundred people carry harmful versions. The rates among certain groups—most notably people with Ashkenazi Jewish heritage—are much higher. About one in forty people with Ashkenazi Jewish heritage carry a harmful mutation. By absolute standards, these numbers may not sound like a lot—in the general population less than a quarter of 1 percent of people have such a genetic alteration—yet this tiny segment accounts for fully 5 to 10 percent of all breast cancer cases and 15 percent of all ovarian cancer cases. It is a disproportionate burden.

It's strange that something so small could cause so much havoc. Cancer is a complicated beast with many different causes, but at their core, they boil down to genetic anomaly. When several important segments of DNA in a single cell get screwed up—by heredity, by carcinogens, by viruses, or by natural mutations—cellular DNA doesn't work correctly, leading to unchecked cell growth, aka cancer. The BRCA1 and BRCA2 genes produce proteins that help suppress the growth of tumors, and they play an important role in repairing cellular DNA. For this reason, when BRCA genes go awry, it's harder for the body to stop tumors from growing, and cellular DNA isn't repaired properly, which may mean it can quickly acquire other mutations that lead to cancer.

However, by itself an inherited mutation in one BRCA gene isn't enough to cause cancer. Leaving aside the chromosomes that determine sex, we have two copies of each gene—one from our mother and one from our father—so if one doesn't work, then the other can fill in. Inherited BRCA mutations raise a person's chance of developing cancer because instead of starting out with two perfectly good genes, BRCA carriers start out with one broken copy. Since as we age we all acquire genetic mutations—the vast majority of which are benign—during cell replication or from environmental exposure, it takes fewer coincidences to muck up a BRCA carrier's single working copy of those genes. Scientists have identified numerous inherited mutations within the BRCA1 and BRCA2 genes that disrupt their tumor-suppressing function and can lead to cancer. The alteration that runs in my family—3600del11—appears to be French.

Though the story of BRCA is partly one of scientific research and medical treatments, researchers racing to unearth the human genome's hidden treasures, and military doctors in combat zones developing the crude tools that will later be used to rebuild BRCA-carrier Angelina Jolie's de-breasted chest, it's also the story of individual women and men choosing knowledge over ignorance and making extraordinarily difficult decisions.

To me, my family's level of cancer feels biblical in scope, and this has mediated my personal experiences with the medical system. As I spoke with various relatives, I began to understand how much this has affected all of us—men and women, whether we carry a BRCA mutation or simply live close to someone who does. Through interviewing strangers who have dealt with the same issues, I also discovered that there is no one defining BRCA experience. Though there are commonalities, every particular family has its own flavor of sadness and triumph.

When I began my research, I did not know that it was possible to give a pregnant woman chemotherapy, but then I talk with a woman diagnosed with cancer while carrying her first child. She informs me that some chemicals cannot break the placental barrier. A man in his sixties describes watching his mother die from cancer when he was a child, and his voice still trembles. He details the guilt of passing a BRCA mutation along to his adult daughter. Shortly after she removes her breasts, he finds a lump in his own. Though it is not cancer, he learns what it means to wait for mammography results. I speak to a woman who watched her father, two aunts, and two cousins die of BRCA-related cancer. She is at the funeral for one of these cousins when the genetic counselor calls to tell her she is positive for the family's BRCA mutation.

Another woman, diagnosed with breast cancer at forty-two, tells me about calling her sister to say she has a BRCA mutation, but when her sister picks up the woman is crying so hard she can barely get the words out. In the years following, all three of her siblings test positive for a mutation. The hardest thing for me to hear, though, is about her young son, who walked in on her once while she was changing clothes. Her mastectomy scars frightened him so much that he fled, crying.

Three sisters with a close relationship test together because their grandmother and several uncles on one side have died of BRCA-related cancer. One sister, a musician and a performer with a cabaret act, tests positive. Another sister tells me, "I remember when I called her to tell her that I was negative and I just burst into tears. And I just said to her, it should have been me, it should have been me. And she said, 'No, I'm glad it is me.'" While the positive sister schedules her mastectomy and hysterectomy, the other sisters care for their mother, who is dying of cancer.

For many families, though, the stories aren't so dramatic. Most of the women with young children that I talk to speak pragmatically, with matter-of-fact voices. They tell me they have chosen mastectomy because they want to be there for their kids. I talk to women who feel the surgeries they had did not change their femininity, or even enhanced it, to women who are grateful for the knowledge, to people who do not want to open these issues and have decided not to test, or who tested but decided against surgery, and to women for whom this knowledge did not evoke a strong emotional response, or at least not one they are willing to share with me.

According to a 2014 study, less than half of the women who receive positive BRCA results elect to have a preventive mastectomy, while three-quarters decide on preventive ovarian removal, though only a fraction of those complete ovarian removal by the recommended age of forty. The rest do not choose to engage in any follow-up, or they try chemoprevention or live with their risk by engaging in surveillance. Typically, the women who choose surgery do so within months of their positive result.

All this, and more, is part of what it means to have a BRCA mutation.

Before their blood travels to technicians at Myriad Genetics—the lab that has somehow managed to patent a piece of the human genetic code—they might imagine the BRCA test as a delicate, sensitive tool, surely one of the apogees of modern scientific development. But positive results, once returned, sealed with their drops of blood, offer women a Faustian choice about whether to risk their lives or sacrifice their breasts and ovaries in the service of continued health. If the test represents the culmination of centuries of research, then the treatments—live with fear of cancer and practice constant vigilance or cut out internal organs—are crude and barbaric.

The ability to test human DNA for susceptibility to disease is miraculous, but it opens up a whole host of quandaries. Does it really benefit us to know what diseases lurk in our DNA? Is it OK to patent a human gene? Can we charge people with genetic cancer risk more for health insurance? Is it ethical or rational to remove healthy body parts? To engineer babies without these predispositions? The shining hope in the bottom of the box is, of course, the ability to out-sneak diseases like cancer, increasing our overall healthiness and life span. BRCA patients and researchers are not the first ones to face these decisions, and they won't be the last, either. As genetic research continues, it's likely that a growing number of people will confront similar quandaries. Already, we know of other genetic mutations linked to breast cancer, as well as mutations linked to bone cancer, leukemia, thyroid cancer, colorectal cancer, eye cancer, kidney and pancreatic cancer, to say nothing of the genes linked to conditions such as diabetes, certain types of arthritis, asthma, early-onset Alzheimer's disease, and more. Perhaps one day we'll simply carry our genetic code around on thumb drives that we'll give to our doctors at check-in.

One thing's for certain: gazing into the crystal ball of genetics has complicated repercussions. It is a bit like the thought experiment of Austrian physicist Erwin Schrödinger, who imagined putting a live cat in a box with a small quantity of some radioactive element. So long as the box is closed, the cat is either dead or alive—until it is observed, in some way it is both; the possibility contains multitudes. Open up the box, and the truth will escape. As with Pandora's fable, you cannot gather up the knowledge or the plagues and stuff them back inside—the tenuousness and bliss of ignorance are gone forever. The only thing left is to cope with the present as best you can.

1 | The Ham Speaks for Itself

The only thing my father remembers about that Christmas is standing in front of the meat case in the supermarket and deciding to buy the whole ham. Ordinarily, the butcher splits a leg of ham in half—the more delicious butt end, from near the pig's hip, and the shank end, each of which sells as an individual ham. At this supermarket, these hams filled the meat case, with a few full-length behemoths skirting the edges, as big as our congealed uncertainty and dread.

That Christmas, my mother received her second cancer diagnosis right before the holiday. As my father tells me years later, "It's sort of like, well, the universe is going to explode, so if I never had a chance to buy a whole ham before . . . what the hell. It's kind of in the same line as 'We may as well go skydiving—what the hell.' I don't remember who else was there or what else we had or anything else."

My uncle Mark doesn't remember the holiday until I mention the food. "Two hams. Yeah, I do. Like forty pounds of ham in the refrigerator. This is one of the things [we talked about]—are we going to finish the hams? Or are they still there? Or how many hams does he think we can have? I just remember eating lots of ham. But again,

I don't remember a lot more beyond that. Your mom's in the hospital; your dad has a lot of ham."

My uncle Alan hardly recalls that Christmas either—it was dwarfed for him by what came after. But, he says, "I remember the hams." And apparently my father "made a joke about eternity or infinity or something like that is a ham," Alan tells me.

"Eternity is two people and a ham," my father corrects, when I ask him. "On cookery," he says, "I've always had a tendency to go grossly overboard, and this may not be the worst example of it, but it's close to it."

My mother swears it was two hams and a turkey.

Cancer has an incredible ability to warp memories as well as bodies, to intensify emotion to the point where it clouds reality. For me, my mother's operations all blend together into one Technicolored nightmare. And although it is her bout with breast cancer that frightens me most, many of the scary memories of watching her sick and weakened derive from her thyroid cancer treatment, which took place during the ham Christmas. The two experiences—the fear of breast cancer and the pain of watching her suffer thyroid cancer—have become conflated. And while I remember the hospital visits, unlike the rest of my relatives I don't remember the ham at all. I remember holidays where we had turkey, lobster, standing rib roast—once even a goose stuffed with prunes stuffed with foie gras—but as far as I can recall, we've never been a ham family. And since that Christmas, my mother has been unable to enjoy ham because the accumulated level of radiation she's gotten to her neck has damaged her salivary glands and given her permanent dry mouth, making even orange juice too spicy for her to drink unless diluted, making salty meals like ham or my father's eight-course Chinese banquets a thing of the past. Even with the gross synthetic spit in a tube she sometimes uses to relieve her symptoms.

The August before the ham Christmas, the three of us had gone backpacking together in New Hampshire. I was six and carried only my stuffed basset hound, Ralph, in my little knapsack, while my parents squired up sleeping bags, clothes, lunch, and the rest of the necessities. While we were there, my mother developed a rash on her face that didn't recede even after we returned home. As with the proverbial old lady who swallowed the fly, things kept escalating. She visited a dermatologist to treat the rash; he noticed a lump in her neck and suggested she get it checked. Her general practitioner discovered the lump was in her thyroid and suggested an endocrinologist. The endocrinologist tried to aspirate the lump and suggested she see a surgeon. When the surgeon called my mom into a room to deliver the results with yet another doctor, she knew it wasn't good. This lump could be something, they told her, and they'd have to operate to remove her thyroid. While they'd try to stay away from her vocal chords, there was a chance those might be damaged, or even cut, during the surgery.

By that time, it was December. My mother scrambled to get ready for Christmas, when our family planned to entertain her sister Cris and brother-in-law Alan; her brother Mark; her parents, Meg and Roy; and a family of four from down the block. In case her voice went, she recorded a tape of herself reading my favorite books, saying our prayers, and singing my bedtime songs. I listened to it every night. When she arrived home from the hospital on December 22, Aunt Cris and Uncle Alan had already shown up for the holiday festivities. And by then we knew that the lump in her neck was thyroid cancer, her second cancer diagnosis in five years.

For cancer survivors, making it to the five-year mark is a big deal because scientists measure cancer survival in half-decade increments.

For example, let's say you are diagnosed with breast cancer. The cancer is assigned a stage, ranging from I to IV, in increasing degrees of severity. Stage I cancer is small and confined to its area of origin, while stage IV cancer has hitched a ride to other places in the body and set up shop. More than 90 percent of patients diagnosed with stage I or II breast cancer are alive five years later, while less than 30 percent of late-stage patients are. In addition, there are different types of cancer. Some are ninjas—nimble, lethal, and hard to kill—while others blunder around like drunken frat boys—clumsy and slow moving, but still capable of pushing you down a flight of stairs if you don't watch it. In 1982, my mother had been diagnosed with stage II ninja-like breast cancer. Although she was only thirty she had half expected it, given her family history of cancer. But after five years of remission, that the new diagnosis was *thyroid* cancer, and not a metastasis of the breast cancer, came as a shock. The breast cancer had been the fruiting of a genetic defect passed down in our family for generations, though we didn't know it yet—just like the stage III ovarian cancer my grandmother had recently survived. But the thyroid cancer was a sporadic cancer. After all, one in two American men and one in three American women will develop cancer during the course of their lifetimes, and the bulk of these are not hereditary—just like the colon cancer that had plagued my grandpa two years before the ham Christmas, or the bladder and prostate cancer that would dog Uncle Alan twenty years later.

The results of my mother's surgery were not good. Not only was the lump in her neck thyroid cancer, but one of the lymph nodes was positive too. The lymphatic system operates like a sort of centralized subway for cancer—once the disease jumps the turnstile, it can take a train to other places in the body and set up little terrorist cells.

My mother remembers wearing a blue cowl sweater dress at Christmas Eve dinner and later at our church outing. All the kids in our congregation dressed in robes at the early service, sat in the front, and sang carols. As a Sunday school teacher, my mother robed up to chaperone. But because of the surgery, she says, "I couldn't sing at all, and my voice was very weak. And I was tired and I was scared because this had come out of nowhere, and I was really scared about making it through."

What little else we collectively remember from that Christmas—beyond the ham and the cancer, of course—comes from my aunt Cris's diary, which she, like her father, Roy, has kept since childhood. That Christmas "was the only time I've seen your father really upset; your father was completely wiped out. He just looked grim. He just looked really, really grim," she tells me on the phone. But after my mother came home from the hospital, he looked better. I gave a dance recital in a silver dress, her diary says. My mother remembers that frock, an outfit for playing dress-up that she had made out of thin, shiny plastic fabric that frayed badly around one of the seams after I'd worn it a few times. We played games and did puzzles, and according to my aunt's diary, my mother proclaimed it the "best Christmas ever."

When I tell my mother about this twenty-five years later, we laugh. Mom, I say, if you could pick any one Christmas to relive, it'd definitely be that one, right? Oh yes, she says, nothing like a cancer diagnosis and organizing a huge dinner while recovering from surgery to make a holiday really great. But after being in the hospital for so many days following her thyroidectomy, she says, being alive and at home probably did feel like the best thing ever. Of course, her side of the family has a tradition of playing their emotions close and their medical diagnoses closer, of smiling and soldiering on even in the face of tragedy. Perhaps this is why I did not know, until I began interviewing my aunt, that the ham Christmas was also my aunt's last one with her natural breasts.

A few days after the holiday, after she and Uncle Alan drove back to their place in Cleveland along with my uncle Mark, she had both of her healthy breasts removed and reconstructed with fat from her abdomen in order to reduce her risk of breast cancer, a cancer that had already hit four members of our family—including her mother and her younger sister.

Neither Cris nor Alan or Mark remembers much of her operation and recovery. Alan and Mark killed time during the eight-hour procedure buying some piece of stereo equipment. Mark wanted to be useful to his sister, in part because he hadn't been too involved in helping out after the cancers of my mom and grandmother, and he felt a bit guilty. On New Year's Eve, Mark and Alan left Cris's bedside to eat oysters at a nice restaurant down the block from the hospital. And eventually, on a cold January day after Mark left, Alan brought his wife home from the hospital, driving carefully in the ice because every bump caused her pain. Since then, my aunt has not been able to sit up from a supine position without the use of her arms, a side effect of the reconstructive surgery that took fat and muscle from her stomach to make her new breast mounds. As the doctors warned her, no more sit-ups.

Meanwhile, back in Washington, DC, in the post-holiday, ham-leftover eternity, things at my house took a sharp turn for the worse. About a week after my mom's thyroid operation, she developed a huge lump at the surgical site. I remember it as a lemon-sized goiter on the right side of her delicate neck. My mother, though, tells me that it was a lump the size of an olive, a lump that felt heavy for its small size, as heavy as her anger and fear and desperation, a solid blockage against her fierce desire to live and see me graduate from high school and college, get married, and live the happy and fulfilling cancer-free life she had planned for me. Perhaps it is a false memory, but I can see her sitting on the beige-carpeted stairs in our living room, touching her neck, saying to my father, “Here, Dickie, feel this,” and guiding his fingertips to it, then laughing hysterically, because of course it wasn’t a lemon or an olive. It was the cancer, come back like a horror-movie zombie that just. Won’t. Die.

“It was at the point where that lump came back that I got really scared,” my dad told me years later. “I guess that thyroid cancer comes, like all these other cancers, in different flavors, and one flavor kills you immediately, and the other flavors are not so serious. So when that thing came back I thought, ‘Oh my God, what’s going to happen now?’ I thought that she was not going to survive this one.”

My mother has had too many hospitalizations. Of course, I can’t possibly remember any of the early ones. I was only eighteen months old at her breast cancer diagnosis, when she had her first mastectomy. Friends took care of me during her chemo and radiation, although once she couldn’t find someone and took me with her to an appointment. The nurses apparently let me watch her on one of the machines through the window, making my mother livid. Close to Thanksgiving, she’d been hospitalized because the radiation and chemotherapy had deprived her body of its ability to fight infection, and she’d gotten a bad hangnail that turned into an extremely high fever. We had been planning to visit her parents in Arkansas, she tells me, so my grandparents simply loaded the turkey into the car and drove to see us in Dallas, where we lived at the time. They brought pumpkin pie to the hospital for her, she said, and the day after she got out we all went on an excursion to the mall so that I could have my picture taken with Santa, a trip that must have exhausted her as much as it delighted me. A few months after her mastectomy, she found a lump in her other breast. “They were pretty sure it was benign, but I thought, ‘Well, if this is happening, do I really need this breast?’ ” she tells me. By the following Christmas, she’d had her remaining breast removed prophylactically, and then a year later she had reconstructive surgery, followed by a hospitalization for some post-surgery complications, and then again for nipple reconstruction with the darker skin from her groin. By that time I must have been four or five.

My mother remembers my fascination with hospital food during one of the visits. The compartments of gelatinous foodstuff repelled her but entranced me. The food came in a special tray! I sucked needles of ice out of the tinfoil lid of the orange juice cup and the nurses gave me ice cream. All I remember about her reconstructive surgery is that another of my aunts fed me spaghetti and guacamole—two of my favorites—and that I wasn't allowed in to see my mother because I was too young, which upset me. It's funny, because my uncle Mark's only memory of his mother's numerous medical treatments two decades earlier is that after her first mastectomy, the doctors wouldn't let him into the ward to visit. He was eleven and remembers thinking that he was very mature at eleven and that it was ridiculous for them to forbid him from seeing his own mother.

My mother wasn't done after the reconstructive surgery. She'd end up hospitalized for the thyroid removal, radiation treatments for the thyroid cancer, an implant replacement, and, five years later, a hysterectomy, appendectomy, and oophorectomy in one fell swoop.

My mother has many memories of her breast cancer treatment. During the mastectomy, she'd been in the hospital over my father's birthday, and since the hospital wouldn't let kids into rooms, she stuffed her bra with Kleenex so she wouldn't look different to me and then came down to the lobby to see us and to give my father a briefcase she'd purchased weeks before. "All in all I tolerated the chemo pretty well," she wrote to me decades later.

> It was awful to lose my hair so quickly—it came out by the handful one day in the shower. I was not sick to my stomach except for one time. I was tired a lot and worried all the time. My taste for certain food left me but I did not lose weight—in fact I gained

> some weight. Your dad was loving and close to me during this time when I felt so unappealing. He took care of you a lot. I am sad to know I just can't remember your second birthday. Sometimes you would put on my wig and go running into our bedroom laughing and we laughed with you.

She remembers my father and me singing "Zip-a-Dee-Doo-Dah" once in the car during the chemo period. "You both were smiling so hard I thought to myself, 'I will remember this time forever,'" she wrote. She wore one of my father's undershirts beneath her gown when they went to the law firm Christmas party, because doctors had drawn on her skin in permanent marker to target the beams for radiation therapy and she didn't want the ink to rub off on her dress. Friends and family called and visited and took care of me while my mother had her treatment. "I had joined a church that I liked and my faith in God also helped—I felt that presence in the many people who helped me," she wrote, from the kindness of a nurse who once gave her a back rub to an old friend who dressed up in nice clothes every single day to visit her at the hospital.

My archetypal memory is this: my mother half inclined on a hospital bed, surrounded with large loud machines and hooked to an IV bag that sends tubes into her hand, where white tape covers the needles. She looks dazed, slightly hazy, but she has spackled a vague open-mouthed smile onto her face for my father and me. Her lips shine with Vaseline that sits in a little paper cup on the bedside table, with a long cotton swab planted in it like a stick of incense. A Styrofoam cup of ice chips sits next to it, the only thing she is allowed to battle dry mouth because solid food might make her nauseous. The skin beside her eyes crinkles because she is pleased to see us, but also because she's in pain. When I am allowed to hug her, if I'm allowed at all, I have to do so gently. The painkiller makes her legs itchy, and she scratches them up and down under the unflattering hospital gown. Sometimes, I watch her sleeping amid the whirrs and beeps of the machines. When her smile becomes too bright and tight, or the hollows under her eyes too desperate, my father and I know it's time to go home.

After the ham Christmas, she received the first of a lifetime of radiation treatments. The thyroid thrives on iodine, so one basic method of killing thyroid cancer is this: carve out what you can, starve whatever remains of iodine for a while, and then feed the patient radioactive iodine. The hungry homegrown terrorists gobble it up and die from the radiation damage. Because of my mom's neck growth, the doctors didn't wait to starve the cells. They started the radioactive iodine immediately. These hospitalizations were the worst because the doctors would isolate her in a corner room and forbid us from touching her or even getting close to her. Too much exposure to radiation causes cancer, of course, so these measures existed to protect hospital staff, visitors, and other patients. They lined the floor and walls with dark garbage-bag plastic and taped a line on the ground that visitors could not cross. When the nurses brought food, they would dress up in space outfits, wheel a cart up to the line, and leave. Then my mother would trundle over with her IV and tote the food to her bed. Visiting her under these circumstances felt cruel when I was a child. My father and I would inch right up to the line and tell her how much we loved her. We stood so close—mere yards—but the gulf between us felt vast and intractable. More than anything, I wanted to run across that line and fling myself around her neck to be held and comforted, yet it was impossible. Every five or ten years, my mother still receives this treatment for the incremental levels of cancer left in her body, though she is approaching the lifetime limit for such radiation, a line beyond which her risk of developing certain other cancers will climb.

The times when she received lower doses of radiation for scans that would check for residual cancer cells were stranger still. She'd be sent home to us, where she'd make our food with plastic gloves on, sleep in the guest bed, and use the downstairs bathroom so as not to irradiate us. Close physical comfort—hugs and kisses—were forbidden for a week or two, until the radiation wore off.

After the ham Christmas, while she lived in that garbage-bag-lined corner room for the first time, my mother sent me flowers. They came in a pot shaped like a chocolate ice cream cone, with a metal spoon bearing a matching ceramic ice cream cone handle sticking out of the arrangement. As a twerpy six-year-old, I had felt left out because her illnesses made my mother the center of attention, though she never courted that, and I saw her getting streams of flowers and other presents. I'd made a jealous comment about it, and she sent me flowers from her hospital bed.

Five years later, she gave the eleven-year-old me another gift, a troll doll, before her hysterectomy and oophorectomy. Though it was a routine surgery, given her history of hospitalizations my fear for her existed in outsize proportion to the actual risk. After the doctors found a benign ovarian cyst, she had decided to get it over with rather than wait around for the ovarian cancer that had killed her aunt and almost killed her mother. Ovarian cancer is one of the most lethal cancers out there—killing more than half the women diagnosed with it—possibly because it is so difficult to detect.

What I remember most about my mother's thyroid cancer, though, is not the ham or the radiation visits or even the flowers. It is making Chinese duck stew with my father—visiting the fancy grocery store to pick out the nicest vegetables, skimming the pot, even wrapping shrimp dumplings—my favorite!—in little wonton pouches to simmer in it. It's curious, as if my mind is trying to protect me from my own memories. I try to recall the cancer, but all I can see is the food.

Her illness and frequent hospital visits had their silent repercussions on me. On some innate level I understood that she could be taken from me at any time; I understood Death as our constant silent companion, someone who could interpolate himself between us and our loved ones on a whim. Like all kids, I could be a jerk sometimes, and of course my mother made me mad, but I couldn't express this anger toward someone who might die. What if those were the last words I'd say to her? Saying "I love you" before any trip—even one to the store—was the norm in our household.

Although her illnesses frightened me, my parents gave me a happy childhood. Outside of the fear that she would die and seven years of psychotherapy, there were ballet practice or gymnastics practice or swimming practice, recreational choruses, Rollerblading with my friends, trumpet lessons, Nancy Drew novels to read, classic novels to be read to me, intriguing clubhouses to form with neighborhood kids down in the wood, wild raspberries to gather in the summertime, hills to sled in winter, Girl Scout camping trips, church picnics, origami animals to fold into existence, and string figures to tangle my fingers in between trying to coax our ancient computer into loading modern adventure games. And of course, a whole string of cooking lessons in desserts and fast meals from my mother, and in breakfast and epic banquets from my father, who seemed to go through culinary phases. We boiled two dozen eggs for different intervals of time to discover the perfect soft-boiled recipe; we ate nothing but sautéed chicken for eight weekends straight until we'd perfected it; we poached fish to exhaustion, scoured supermarkets for contraband spices, carved turnips into football shapes, and flipped crepes.

But my mother's illness had already lodged beneath my skin, and I grappled with big questions of mortality and life planning. Around age ten, a few years after the ham Christmas, I decided I'd rather not die. But since that seemed a biological impossibility, I decided I wanted to die while doing the things I loved. So, in other words, doing cartwheels in a pool while singing after a taco dinner. By the time I turned twelve, I'd had years to consider what not-dying would look like, and I'd decided it meant having a legacy that lived on through the arts, preferably screen acting or virtuoso trumpet performance. The arts are a foolish career choice, but then, if you are going to die of cancer at thirty, you don't have much time to suffer in poverty and obscurity—the world is ending soon anyway, so what the hell? In junior high, I mapped out the rest of my life. If I was only guaranteed existence until the age when my mother first got cancer—and if that was not a certainty, it felt at least a firm possibility given my family history—then I had to meet the man I would marry by age twenty-four at the latest, so we could date for a year or two and still have time to pop out some babies before I croaked. I was determined that disease and illness would not deprive me of anything I felt entitled to—love, a family, a career. Passing on without a legacy would mean letting the cancer win.

After I graduated from high school, my mother and I spent a month road-tripping across the American Southwest, clocking some quality mother-daughter hours together before college separated us semipermanently. Somewhere near the Grand Canyon, while we waited for a shuttle or bus or possibly for my father—who joined us late in the trip—to fetch the car, we talked about cancer. It hadn't been a forbidden topic or anything—I remembered my mother's cancer, obviously, and when I was a little nibbler Aunt Cris decided not to have children, so she passed along a silver daisy ring to me, the ring my great-aunt Trudy wore until she died of breast cancer at age thirty-one. It's one of the few family heirlooms I possess, and I wear it always. Since cancer has tortured my family across several generations and there is so much accumulated pain, my mother and I seem to talk about it in uncontrolled spurts. The topic comes up, like a tulip bulb peeking out of the soil, and we make temporary space for the whole conversation. That night was one of those times. Sitting on the park bench at dusk, smelling the dry piney air and looking at the smooth beige stones composing the pavement, we talked. Though it was not the first time I had gotten this information, for the first time, I began to really understand. Things were far worse than I imagined. It was not just my mother, but my grandmother, and her sisters, and their children. Every female relative on my mom's side of the family had lost her breasts to cancer or the fear of cancer or both. It dawned on me that the story wasn't over. I was also a female relative—the oldest of my generation in the extended family. We left the conclusion unspoken. Neither of us was prepared to face it just yet, and since I was only seventeen, the decision was probably years off.

By the time I reached my late teens, a graph had imprinted itself in the back of my mind, an image of what life is like, age plotted against awesomeness, which resided on the y-axis. The graph described a rising curve that peaked around age twenty-one before dropping off sharply. At thirty, or a little later, the chart ended. At the moment of our talk, living near the apex of the curve, I could see the downslope coming already. I had spent some time considering what I'd most regret if I died tomorrow. At that age, I felt a vast gulf of hipness separated me from my parents, and I suspected that since they were kind, responsible, reasonable people, they had never truly lived. I didn't want to be like them. I wanted more recklessness in my life. If I died tomorrow, I'd regret never having stayed up till dawn dancing, never trying steak tartare or sucking down a few slugs of authentic absinthe. I'd regret saying no to offers to try sea slug, visit Oslo, go on an archaeological dig. I'd regret never having known the sensation of a shaved head or a tattoo needle. I'd regret letting fear of failure hold me back from blundering into places where I didn't belong, putting on performances, writing bad stories. Most of all, I'd regret not having done everything within reason to pursue my dreams.

This feeling that life is guaranteed only until the date of your mother's first cancer diagnosis has infected other relatives as well. My great-aunt El—one of my grandmother's older sisters—had breast cancer in her thirties when her four kids were young. Though they are more than twenty years older than me, I share this experience of a life mapped out with Kathy and Lisa, El's two daughters, my mother's cousins. When El was diagnosed with her first cancer, Kathy was about seven, and Lisa was an infant. I was six for my mother's thyroid cancer diagnosis and an infant for her breast cancer diagnosis. We feel developmentally akin. When I talk to Lisa, she admits that cancer "always seemed inevitably that it was to be my fate. It happened to my mom. It was so much a part of our lives. I always thought that some day I'll get breast cancer." Kathy decided that she needed her PhD by age twenty-four so she could work for six years before succumbing to a planned bout of cancer. As she puts it, after watching her mother sicken as a child and again as an adult:

> I never developed that illusion that you're guaranteed a long life. There are people who walk around who have this lightness of being that of course they're going to be old and they'll make jokes about it. There are people who never had that in them and I never had that. . . . I remember sitting in California on a dock with my husband (the right one). We'd been married two years, and I didn't realize it but I was actually pregnant at the time, and he said, 'OK, we're going to do something you've never done; we're going to talk about the future.' And he said we'll make plans about the future, and he said that you have to think that we're going to live to be old. Because I never talked about the future—I didn't expect it. I think that's part of why I raced through academics, because all you were guaranteed was the day you have. . . . I never saw my mother old. I didn't know what I'd look like as an old woman. I have girlfriends who talk about plastic surgery and I'm so not there. I'd love to have wrinkles on my face.

This sense of life as a precious, fleeting commodity perhaps explains my mother's constant activity. "I think I tried very hard to live as though I wouldn't let the diagnosis of cancer interrupt my life—probably my survival strategy, but it also kept me in some ways from asking for help very often," she wrote to me. She packed up her wig and prosthesis and we went on our planned family camping vacation with my father's sister and her family. Although my mother never returned to work after the breast cancer, she filled her hours with volunteer activities. They've changed over the years, but here's a sampling of a month for my mother: Every week she volunteers at a local arboretum where she gardens, and in a reading-intensive first-grade classroom at a public school; she also gardens for her church, where she taught Sunday school for twenty-five years. Then there are the book club and the museum Wednesday club she organizes once a month and the quarterly trip to the homeless shelter to make dinner. She walks four miles every day with a friend from down the block, gardens her own garden, lifts weights, and then there are the needlepoint and sewing projects. She can't even sit still to watch TV.

In my early twenties, I stopped worrying so much about my mother, and slowly, inexorably, my fears for her transferred over to me. At eighteen I met the man I would eventually marry, but we didn't start seeing each other for several years. I traveled places. I learned what it meant to be drunk at a dance club. A woman in a park in Barcelona, where I went exploring with friends, handed me a sprig of rosemary, touched my palm, and predicted that I would have lots of health problems. Though I don't believe in fortune-telling, in the years to come the scene replayed in my mind. I clung to my friends too tightly and messed up many relationships with my clutching. Eventually, my future husband and I settled in Boston, in a 1950s-style apartment over an old dance hall with original linoleum and not enough insulation, and I began grad school. And one night, while looking at the vulnerable pale-blue veins under my breasts in the mirror—a favorite pastime when I felt twee after dark—I decided to do a breast self-exam. I discovered a weird thickening on the underside of my left breast. George, my boyfriend, could feel it too. Intellectually, statistically, I knew it was probably nothing, but I booked an appointment, and yep, the gynecologist found a lump. She sent me for a mammogram and an ultrasound. I was more scared than I was willing to admit, and at the doctor's office I kept waiting for someone to tell me "You're fine," or "It looks all right." As it turns out, the technicians are not permitted to offer these assurances for fear they might raise hopes that later turn out false.

The thing in my breast was nothing. I was twenty-three and still felt invulnerable. But my doctor was concerned. Some experts recommended starting mammograms ten years before the diagnosis date of your closest relative with cancer, she said. For me, that would have been age twenty-one. I was already behind.

A year later, my mother brought up genetic counseling during a phone call. I should consider it, she said. I scored a referral to a breast center in Boston, appalled to discover that it shared space with the oncology department, a place I had spent my whole life trying to avoid.

For the last decade my grandmother had devoted herself to finding out what was in the family. She enrolled herself in a genetic study at Creighton University and wrote to German relatives she'd never met or hadn't seen in four decades to assemble a larger pedigree. She had tested positive for a cancer-causing genetic mutation as part of the study, she said, and we mostly believed her; but she was also in decline both mentally and physically, and my mother didn't want to worry her by bringing up the topic again since she'd had enough cancer in her lifetime—three bouts of her own and eight bouts in close family. So it made sense to test my mother first because she had already had premenopausal estrogen-receptor negative breast cancer, that is, cancer that struck before menopause and that did not contain receptors that made it grow in the presence of estrogen. Both the premenopausal and estrogen-negative elements are risk factors for a BRCA mutation. Once the technicians at Myriad Genetics had sequenced her DNA for a several-thousand-dollar fee covered by insurance, all they would have to do would be look at the spot in my DNA where my mother's mutation, if any, occurred—a cheaper procedure. If my mother tested negative, I would be off the hook unless we had some alteration science had not yet discovered. If she tested positive, then there would be a fifty-fifty chance I also carried the genetic mutation and was at grossly elevated risk for developing breast and ovarian cancer at an unusually young age.

At twenty-five, I no longer felt invulnerable. I'd recently had my first serious interaction with the medical system: a retinal detachment in each eye, spaced a year apart. The retina is the membrane that coats the inside of your eyeball and contains the rod and cone cells that detect light and color, and permit vision. Since I am nearsighted, my eyeball is elongated, which can put extra stress on the retinal lining. Sometimes, a hole develops, and the vitreous humor that fills the eye seeps behind, peeling the whole thing off like wet wallpaper. When that happened to me, I saw warped lines of text, and cigarette burns, the kind that join strips of film together for theatrical showings, black flashing bubbles in the lower inner quadrant of my vision. It's unusual in such a young person, the ophthalmologists told me, but sometimes these things do happen. A retinal detachment is an emergency. If left untreated, you go blind. But since going blind is not life threatening, the procedure to fix it is considered "elective." So one morning I was at the eye doctor thinking I had an infection that some drops would fix, and by the evening I was at the hospital, where surgeons sewed a silicone band to my eyeball while I was sedated but still awake. It wasn't pleasant, but it was better than being blind, I kept telling myself. Afterward, I felt fragile, the unlucky winner of some shitty lottery. Maybe this is what I get instead of the gene, though, I thought. Maybe I had lucked out.

It felt like an eternity—or a few hams—away when I was fifteen and planning the rest of my life, but now, at twenty-six, I began to realize that thirty was not that old. I stopped remembering how tough my mother's sickness felt for me, and I started envisioning what it was like for her. Some of my friends had begun to have children. I imagined them stricken, trying to juggle chemotherapy and careers, trying to lift their babies with arms weakened from surgery. I imagined my beautiful friends, for whom fashion could be both a pleasure and a chore, I imagined their shiny hair falling out, their eyebrows no longer waxable, their faces sunken or swollen from steroids, their turbans, their sense of grimly soldiering on despite the blows to their femininity, and slowly my image of them lost color to become the black and white of old photos, of the faceless horde of women who came before me. This wasn't a fantasy, it was the reality lived by my mother, grandmother, great-aunts, and maybe, in the distant mirror of the future, me.

2 | "It's Everywhere"

Fear of cancer is an old emotion, because the disease is ancient.

Paleopathologist Arthur Aufderheide, a distant cousin of my father, performed autopsies on mummies. In 1990, while on an archaeological dig in the Peruvian Andes, he sliced open a woman about my age, in her mid-thirties, but a thousand years dead and mummified by the environment. Beneath papery dried skin on her upper left arm, he found a knob of bone cancer. He wasn't the only paleopathologist to find this sort of thing in mummies—cancerous abdominal tissue discovered in mummies in Dakhleh, Egypt, dates to 400 BCE, for example. Sometimes, archaeologists don't find actual cancerous tissue but rather the signs that cancer has traveled there before, such as the tiny holes bored in bone by advanced strains of skin or breast cancer. Anthropologist Louis Leakey found a two-million-year-old jawbone that contained signs of a rare type of lymphoma, although pathology never confirmed the diagnosis.

The oldest written record of cancer—the Smith Papyrus—describes forty-eight types of malady with scholarly detachment, everything from broken hands and crushed vertebrae to skin abscesses, and includes both palliative and curative treatments.

Although written in the seventeenth century BCE, it's believed to contain teachings from a thousand years earlier from the Egyptian doctor Imhotep. Case forty-five involves "swellings on the breast, large, spreading, and hard; touching them is like touching a ball of wrappings, or they may be compared to the unripe hemat fruit, which is cool and hard to the touch." In his book *The Emperor of All Maladies,* oncologist and cancer historian Siddhartha Mukherjee recognizes the words. There could "hardly be a more vivid description of breast cancer," he writes. Yet Imhotep prescribes no poultices or balms for this case. The section under "therapy" reads, "There is none."

The semi-mythical Greek physician Hippocrates named this inexorable illness *karkinos*, meaning "crab," around 400 BCE. It's unclear exactly how Hippocrates meant this crab metaphor—tumors can be hard, like a crab's shell, or perhaps cancer spreads like a crab on the move, or causes pain like a crab pinch. Later Roman doctors adopted the name *cancer*, the Latin word for crab, and extended the metaphor. Surrounded by filaments—probably blood vessels—some tumors resembled crabs, their legs reaching into the surrounding tissue.

When cancer goes untreated, tumors become painful, burst through the skin, and exude smelly dark liquid. Because breast cancer is one of the most common cancers—in our disease-prone day and age it will afflict one in eight women—and one of the most visible cancers, occurring in a part of the body where it may be felt, and sometimes seen, rather than in a more hidden internal organ, historical records have no shortage of women who have suffered and died from the disease. Atossa, a queen of ancient Persia and wife of the ruler Darius, let a slave cut into her breast to lance her tumor, her survival probably proving that it was not cancer, but a benign breast disease. Louis XIV's wife, Anne of Austria, had been a fastidious woman in life, but as she died the servants who changed her bandages covered their noses with perfumed handkerchiefs to escape the putrid stench of her weeping tumors. It must have been humiliating.

Before he became a monster, Adolf Hitler was a momma's boy. His mother, Klara, developed a breast lump in 1905 but did not mention it to her doctor until several years later, when the chest pains began keeping her up at night. A mastectomy was not enough to save her, and Hitler went begging to the family's Jewish doctor, Eduard Bloch, for something—anything—that might prolong her life. With the usual options exhausted, Bloch offered an experimental and primitive chemotherapy treatment. He reopened Klara's mastectomy scars and applied an iodine-based medicine that burned its way into the tissues. Each treatment left Klara screaming and in pain for hours afterward, and as a side effect, paralyzed her throat. For forty-six days she weathered the ordeal while her son watched and cared for her. Though Klara died, Hitler felt so grateful to Bloch that he paid the doctor in full and sent him yearly Christmas cards for a time afterward. Thirty years later, Bloch used his connection with the family to save himself, secretly contacting the Führer's younger sister to speak on Bloch's behalf. Within weeks he had all the paperwork he needed to emigrate to the United States, as well as an exemption from wearing the *J* on his clothing. As breast cancer historian James Stuart Olson put it, "It was a curious irony: while Hitler contemplated the liquidation of millions of Jews, he made sure one escaped."

Where cancer goes and medicine fails, fear follows. My family knows this fear; in some ways we've made it our friend, used it to spur us on to medical treatments that other people—mostly those who don't understand the history of cancer in our family—find a bit extreme. Our suffering begins with Gertrude Frida Muehleisen, my great-aunt Trudy, whose ring I wear. Her story, passed down through generations clouded with their own anxieties about cancer, reads vaguely like a Rorschach inkblot for the unifying experience of cancer: fear.

Though she has a very German name, Trudy was born in China on April 23, 1921, to my newly wed great-grandparents, missionaries for a denomination that would later become the United Church of Christ. Though my grandmother's loopy cursive leaves so much out, it notes Trudy's baptism, presumably at the hand of her father, thirteen days after her birth. A younger sister, Elfrieda Katherine, followed in 1922. At the time, kids of missionaries usually got packed off to boarding school. My great-grandmother, whose parents had been missionaries in India, had hated being separated from them as a child and wanted a different experience for her own children, so when Trudy was two and Elfrieda—El—was one, the family packed up and moved to the United States, eventually settling near Milwaukee. Five years later, in 1928, my grandma Margaret—Meg—became their first and only child born in the United States.

They arrived just in time for the Great Depression, a tough time for most Americans and tougher still for pastors like my great-grandpa, who relied on his congregation's donations to survive. Perhaps this explains why so many of my grandmother's childhood stories revolved around food—barrels of sauerkraut her mother cured on the porch, memories of parishioners buying two heads of cabbage and donating the wormy one to her family, her prodigious love of hard-boiled eggs, and occasions when she was dolled up for her father's visits to congregants during which she maneuvered for treats with backhanded suggestions like, "If someone offered me a cookie, I wouldn't say no."

By all accounts, my great-grandmother was a difficult, critical woman who did not bear the family's new poverty well. She also suffered from short-term depression and psychosis, if the genealogy notes my grandmother left behind are correct, and this affected her three daughters. And something about the era, with its paucity of resources, or her personality, seemed to make her daughters competitive with one another. If you ask El's family, they'll say my grandmother Meg was the favorite child, the youngest, the baby, the daddy's girl. If you ask my family, they'll say that Elfrieda, the great beauty, the brainiac, was the favorite. Trudy has no family to speak for her, so her position remains a mystery.

One thing is clear, though: El and Trudy couldn't wait to leave. Both obtained nursing degrees in their mid-teens, moved out, and began working. Trudy fled far, to Newfoundland, Canada, and followed the family's humanitarian bent, nursing the indigent poor at the Grenfell Mission, one of the region's first permanent medical facilities. For a few years, my grandfather says, she did everything from milking cows to delivering babies. She met, fell in love with, and became engaged to a young man—no one remembers his name—who met a tragic end. He took a boat out into the ocean, and the boat came back without him. My grandmother believed that her eldest sister never got over this loss, that it explained part of what happened afterward. My grandpa thought it contributed to Trudy's somber nature—she was less vivacious and feisty than her two younger sisters. Eventually, Trudy moved back to Milwaukee and worked for the county as a nurse, paying house calls on locals.

It's not easy to talk about cancer in the best circumstances, but in the 1940s and 1950s, social norms made it even harder. As my grandfather puts it, "First of all, talking about breasts was pretty much off limits, and talking about breasts and having a problem with them was even more secretive. You'd tell someone you had a broken leg, but you didn't want to tell them you had stomach cancer, liver cancer, breast cancer, cancer of the brain. Those were always way out of the limits of public conversation." The titillation of breasts, the secrecy that shrouded cancer, and the intense emotions illness evokes muddle the narrative. Perhaps this explains why three different stories have come down to me.

My mother, my aunt Cris, and I all inherited the same version: Trudy knew she had cancer and chose to do nothing about it because the only treatment available—a radical mastectomy—terrified her. The story carries the intimation that she struggled with the loss of her fiancé and saw no reason to live. As Cris puts it to me, "She sort of existed as this tragic romantic character in the background." The tragedy and romance appealed to me as a young woman too—in college I'd write my first real short story about Trudy. For me, the moral of this story is that Trudy was smart enough to be terrified of the treatment but not brave enough to make the brutal decision to take off her breasts.

El's eldest child, Kathy, inherited an altered version of the tale. Trudy first felt the lump in her breast at age twenty-seven or twenty-eight, but as a nurse she either assumed she was too young to get cancer or she didn't want to undergo a disfiguring operation because she was single and despaired of getting a husband. So she watched the lump. By the time she did opt for a mastectomy—contrary to my family's version—the cancer had spread. For Kathy, the moral of the story is "Do not delay."

My grandpa provides yet another account. After Trudy's mastectomy, he says, she did better before taking a sharp turn for the worse. He and my grandmother took her in because she was single, and later my great-grandparents moved in as well. As Trudy sickened, the family became more desperate, eventually pinning their hopes on the Lincoln cure, which involved injecting, sniffing, or topically applying bacteria that had been destroyed with viruses. The so-called cure, invented by Massachusetts physician Robert E. Lincoln, who eventually lost his license, was used to treat everything from sinus trouble to cancer. Although my grandfather thought it was a scam, he drove Trudy down to the clinic for her treatment at least once. Even if Trudy was strong enough to bear up to the mastectomy, in this version she falls victim to chicanery.

No matter whether she succumbed to fear, delays, vanity, or false hope, Trudy's story is cautionary. All versions end the same way: Trudy died of breast cancer in 1953 in my grandparents' home in Wisconsin at age thirty-one, leaving behind bereft parents, bereft sisters, and five small nieces and nephews—my aunt Cris and my mother Gretchen, and their three cousins. Everyone mourned her rotten luck and lost life, of course, but at the time it seemed like an isolated incident.

If Trudy's tale is cautionary, then El's is historical, encapsulating many developments in modern medicine. Like Trudy, El fled home in her mid-teens to become a nurse. After receiving her nursing degree she wanted to go to university, so she cut a deal with Beloit College in Wisconsin: she'd be the nurse in exchange for tuition, room, and board. While she was there she cared for a soldier who had contracted malaria in the Philippines during World War II and who had returned to Beloit to finish his degree. Ralph's medical release required him to seek regular care from a local medical practitioner. The army probably meant only for a few months, but he did the army one better and married her, not knowing that, by the end, he'd be the one caring for her. They moved to Rockford, Illinois, where he sold large factory apparatus and she defied traditional expectations by pursuing her own career outside the home, becoming a teacher at a local nursing hospital. She had four children, two boys and two girls in quick, alternating succession, giving birth to her youngest, a daughter, in January of 1957, just four years after Trudy's death. That September, she found a lump in her breast and was diagnosed with breast cancer. She was only thirty-four.

Luckily for El, in between Imhotep's dire pronouncement about cancer treatment and her own diagnosis, science had come a long way. For millennia, the concept of humors had dominated Western medicine. Hippocrates—the eponymous source of the oath to do no harm that doctors still take today—believed that the four ancient elements of air, fire, water, and earth had analogues in the human body in the form of blood, yellow bile, phlegm, and black bile. His theory held that healthy humans maintain a balance of the four fluids, but when they get out of whack, illness results. The writer and Roman physician Galen, who practiced around 180 CE, applied humoral theory directly to particular illnesses: cancer, for example, resulted from an excess of black bile.

Galenic theory treated illness as a systemic problem that required a systemic, not a local, solution. By that reasoning, the surgical removal of a tumor might bring immediate relief but failed to address the underlying cause because it did nothing to restore the correct balance of humors. Due to Galen's influence, doctors largely abandoned the idea of surgery as a cure for cancer in favor of systemic bleedings and purgings. The anatomists who mapped the human body from the 1500s to the late 1700s—often dissecting corpses nabbed from graves or the thieves' gallows—upended Galenic theory. Their failure to find any trace of black bile in the normal human body laid the groundwork for surgical extraction of tumors. If illness wasn't due to humoral imbalances, then perhaps cancer wasn't a systemic disease, as Galen would have it. Perhaps it was local and could be cured through the knife.

Until the late 1890s, surgery was a gruesome, life-threatening procedure, often performed without the mid-century development of anesthesia and the later spread of antisepsis. Surgeons executed their jobs in coats spattered with festering body fluids from earlier operations—stained jackets meant a doctor had a lot of experience hacking patients open; stains were badges of honor. In an era before doctors understood germs as the vectors for disease, many patients died of postoperative infection.

The development of anesthesia, cellular pathology, and germ theory in the mid-1800s revolutionized the medical landscape and made complex surgery—including surgery for breast cancer—possible and survivable. In 1847, Hungarian-born obstetrician Ignaz Semmelweis unearthed the startling solution to the problem of infant mortality at the University of Vienna. The university hospital had two maternity wards, each of which delivered about thirty-five hundred babies annually. Doctors and medical students staffed one of the wards, while midwives staffed the other. In the doctors' ward, about 20 percent of the mothers—around seven hundred women per year—died. Meanwhile, the midwives' ward sustained a maternal mortality rate of only 1.7 percent, or about sixty women. Semmelweis decided to investigate that striking difference. He noticed that the doctors did something the midwives didn't—they often performed autopsies in one section of the hospital and then hustled over to the maternity ward to deliver babies. Perhaps, Semmelweis mused, the doctors were transferring some invisible illness from the corpses to the mothers. He began washing his hands with chlorine solution and got the other doctors to do the same. One month later, the maternal death rate in the doctors' ward had dropped to 1 percent. Scottish surgeon Joseph Lister's research confirmed Semmelweis's observations. Inspired by Louis Pasteur's recent discovery of microbes, Lister began sterilizing his hands, bandages, and his patients' surgical sites with carbolic acid, dropping infection rates drastically. He published his results in 1869 in the *Lancet*, and antisepsis was born.

In addition to contending with post-op infections, surgeons back in the day had to operate really fast. No matter how drunk you get someone, they will still wake up and thrash around if you saw off their leg, and it's pretty hard to do delicate, complex work under those circumstances. As scientists in Europe developed antiseptic techniques, surgeons in America discovered the wonders of anesthesia. After months of dosing cats and dogs with ether, dentist William Thomas Green Morton moved up to humans who needed teeth extracted. On October 16, 1846, at Massachusetts General Hospital in Boston, he knocked out a patient for half an hour while physician John C. Warren removed a facial tumor.

Though the twin discoveries of anesthesia and antisepsis spread slowly, they gradually made surgery less painful as well as more survivable. At the same time, advances in cellular theory laid the groundwork for modern methods of cancer diagnosis. In the late 1830s, German botanist Matthias Schleiden and physiologist Theodor Schwann discovered that all living things are made out of cells. German pathologist Rudolf Virchow built on this idea, theorizing that cells only came out of other cells. This had direct implications for cancer medicine. If true, it meant that living things grew either when their cells got bigger or when their cells multiplied, and Virchow extended these observations to every human tissue. By examining cancers of bone and connective tissue under a microscope, he figured out that cancer was a disease of unchecked cell growth. Virchow found that the nuclei of normal cells looked different from their cancerous brethren. By the time he died in 1902, medicine understood that, at core, all cancer arises from excessive cell growth and that malignant lesions and benign ones look different under a microscope.

Enter the epic surgeon William Halsted, a man of outsized importance and big ego whose name would be forever attached to the radical surgery that mutilated and saved both El and my grandma Meg, and scores of other women over the two generations that his influence lasted.

Born in 1852 to a wealthy family, William Halsted was a flawed, complex man. Known as a party boy and athlete in his youth, he graduated from both Andover and Yale before hunkering down as a medical student at the College of Physicians and Surgeons in New York City in 1874. Here, his nervous, obsessive personality manifested through his devotion to anatomy; he devoured the textbooks. By the time he became a surgical resident at Bellevue Hospital in New York City, he'd already had his first nervous breakdown.

Halsted made his mark on the medical profession in part because he appeared at just the right time. Advances such as antisepsis had been discovered in Europe but had not spread to, for example, the hospital where Halsted began his career. As Mukherjee writes, he "entered surgery at a transitional moment in its history. Bloodletting, cupping, leaching, and purging were common procedures." In 1877 Halsted left for a two-year tour of Europe, where he met and studied with the continent's best surgeons and pathologists—including Virchow's students. And he brought some of their practices—including antisepsis—home with him.

Halsted pioneered many important medical techniques, and he must have had nerves of steel. In 1882 he removed his own mother's gallbladder on her kitchen table, one of the first such operations; he was also the first surgeon to successfully transfuse blood—into his own sister, who was bleeding after childbirth. An early adopter of antiseptic techniques, he built his own facilities in a tent outside Bellevue because he found the hospital's operating theater filthy. When one of his nurses suffered raw hands from sterilizing them in carbolic acid before operations, Halsted asked the Goodyear Rubber Company to manufacture a rubber glove, and he introduced them into the operating room. In the mid-1880s during the height of his surgical career in New York, Halsted and his colleagues experimented with cocaine in order to prove its use as a local anesthetic, injecting themselves along the major nerves. Ignorant of the drug's addictive powers at first, Halsted ended up a cocaine fiend. A trip to a sanatorium in Rhode Island cured him of his dependency by hooking him on morphine. He'd fight the addiction to both drugs for the rest of his life while still maintaining an extraordinary practice.

Halsted moved with his wife from New York to Baltimore in 1889, where he joined the brand-new Johns Hopkins Hospital as the surgeon-in-chief, but he retired from social life and became a recluse. His perfectionism and desire for control manifested themselves again. He self-administered morphine on a strict schedule and in tightly controlled doses and raised purebred dachshunds and thoroughbred horses. He bought enough French shirts that he could send them back to Paris to be laundered because Americans couldn't starch a shirt to his requirements. With the same precision and fanaticism, he devoted himself to curing breast cancer.

Now that anesthesia had made long operations possible and antisepsis had made surgery less lethal, surgeons could perform successful mastectomies more frequently. But patients regularly relapsed months or years after surgery, often around the margins of the original operation, as if shards of cancer had been left behind. English surgeon Charles Moore noticed this pattern in the 1860s and solved the problem by taking more tissue; to spare women by removing less of their bodies was a "mistaken kindness," he suggested. German surgeon Richard von Volkmann removed a minor pectoral muscle along with the breasts and experienced a recurrence rate of "only" 60 percent. Halsted thought he could do better. So he pioneered a more extensive mastectomy that would become the gold standard for breast cancer patients for the next two generations.

New York surgeon Willy Meyer and Halsted independently arrived at the same conclusion in the 1890s—the pectoralis major as well as the minor should be removed. Halsted dubbed the procedure a "radical mastectomy," meaning "radical" in its original Latin sense of "root"—he intended to carve out the roots of breast cancer. But soon, the phrase would also come to mean "extreme." Removing the breast and pectoral muscles wasn't enough for Halsted. He knew cancer was a cellular disease and worried that surgeons who cut into the breast or used their hands to remove tissue might spread cancer cells elsewhere in the chest cavity. By the turn of the century, Halsted advocated an operation that removed breast, pectoral muscles, and lymph nodes in the armpit in a single block.

The brutal procedure caved in women's shoulders and harmed their arm mobility. Removing lymph nodes could cause patients' arms to swell painfully with accumulated fluid, a condition known as lymphedema. But the practice also saved lives. And in advanced cancer patients, the operation was palliative—delivering them from tumors that might otherwise burst through their skin—preventing them from rotting to death. Halsted presented a stunning paper on his patients at the 1898 American Surgical Association meeting in New Orleans. Of 133 patients, 76 were more than three years out from surgery, and 52 percent of them had not relapsed—a significant 8 percent improvement over Volkmann's less radical procedure. In the coming years, Halsted performed hundreds of mastectomies, gathering data that allowed him to assign stages to tumors and helped with prognosis. He figured out that cancer that had spread to lymph nodes had a poorer prognosis than cancer confined to the breast, and he concluded that earlier surgery, ideally performed before cancer spread to the lymph nodes, worked better. As Mukherjee points out, Halsted's success kicked off a "macabre marathon" to see who could carve out the most tissue. Surgeons cut through the collarbone and further into the chest to excise the lymph nodes stationed there.

The surgery undoubtedly saved lives, but it relied on a faulty one-size-fits-all approach to cancer and on a misconception of cancer as a local disease that spreads slowly outward from the site of origin. As we now know, some cancers grow slowly, and others grow quickly, or simply aggressively, hitching a ride to other tissues even when the original tumor is small. Mukherjee makes the point that a radical mastectomy cannot cure a woman with metastatic cancer, cancer that has spread to other sites in the body, because "her cancer is no longer a local problem." In contrast, a Halsted mastectomy might cure a woman with a small local tumor, "but for her, a far less aggressive procedure, a local mastectomy, would have done just as well."

In 1957, thirty-five years after Halsted died, my great-aunt El had his eponymous mastectomy on one of her breasts. She had barely reclaimed her thirty-four-year-old body from giving birth to Lisa, her youngest, in January, and by September the dreaded diagnosis arrived. In a letter to her mother, El wrote that she would have the mastectomy because she needed to continue living—she had Ralph and the kids to think of—and that she wasn't going to be like Trudy. The procedure removed her breast and pectoral muscles, along with the lymph nodes in her armpits and under her collarbone, and the nodes deep in her chest that were positive for cancer. The doctors also biopsied her collarbone and rib bones.

The surgery wasn't the only treatment that disfigured her. By the time El developed breast cancer, scientists had discovered more of the weapons in the modern cancer-fighting arsenal. Since she worked as a nurse at a teaching hospital, she had many friends—doctors and surgeons—conducting research. Kathy, her oldest daughter, says she received the most up-to-date medical treatment, including the newer cancer-fighting weapons of radiation and chemotherapy. Radiation had been around for a few decades, but chemotherapy was in its infancy at the time. Kathy believes that her mother was part of a study on chemotherapy, but unfortunately, the details of El's treatment have moved into the opaque past—her medical records are long gone.

Radiation therapy and chemotherapy, like so many medical advances, intersect with the legacy of war. In the case of radiation, the intersection is indirect, but personal. The phenomenon of radiation had been discovered in the last years of the nineteenth century. The strange invisible rays could penetrate skin, revealing a person's inner skeletal structure, and they could also harm humans. Marie Curie's hands blackened and peeled after she refined a tiny quantity of radium, for example, while a vial of the element in her husband Pierre's waistcoat pocket scarred his hip. The burns went deeper than just the skin—as scientists later figured out, radiation had the power to damage cellular DNA, which either made cells die or stopped them from dividing. The first use of radiation in cancer treatment occurred in Chicago in 1896, one year after the discovery of X-rays. Medical student Emil Grubbe figured that radiation might kill rapidly dividing cells—cancer—so he rigged up an X-ray tube, and each day for eighteen days he bombarded an elderly woman with advanced breast cancer that had recurred after her mastectomy. Although her breast tumor shrank, her cancer had already metastasized to other parts of the body, and she died a few months later. Still, the treatment was promising. Once the Curies discovered radium—which produces beams of energy more intense than X-rays—oncologists had a more powerful weapon against tumors. Unfortunately, no one understood that the power of radiation was truly awesome. Marie Curie succumbed to leukemia in 1934, brought on by her unshielded handling of radioactive materials. And in 1960 Emil Grubbe died of multiple forms of cancer after enduring many operations to amputate his fingers, left hand, and part of his face due to radiation damage. Radiation could shrink tumors, but too much of it caused cancer and death. Figuring out the right dosage to shrink tumors without causing enduring harm took decades.

Radiation turned doctors into Goldilocks, searching for the dosage that was "just right." Too much radiation had nasty side effects and could kill healthy organs and normal tissue. Too little radiation failed to shrink tumors. Geoffrey Keynes, the younger brother of the influential economist John Maynard Keynes, played an instrumental role in promoting radiation as a form of cancer therapy. After a stint as a medic during World War I, Keynes decided he'd seen enough dismembered body parts for several lifetimes, so surgical solutions to cancer—and the Halsted mastectomy in particular—repelled him. He turned to radium as a possible therapy for breast cancer. At first, he tried placing radium pellets into large inoperable breast tumors and found that this shrank the cancers. Next he began to wonder whether radiation, used in combination with conservative breast surgery—lumpectomies, which removed only the tumor and a little surrounding tissue, or simple mastectomies that did not take chest muscles—might cure women with less advanced cancers. In 1935 he compared the survival rates for patients treated with surgery and radiation at his hospital—St. Bartholomew's in London—to patients treated with radical mastectomy; it was nearly identical at five, ten, and fifteen years out. Forty-two percent of both sets of patients were alive a decade and a half after treatment. Conservative surgery plus radiation worked just as well as Halsted mastectomies. Although Keynes's startling discovery in England should have shaken the medical establishment, America ignored his research until the mid-1950s. The influence of Halsted's surgical radicalism held such sway that lumpectomy plus radiation didn't gain mainstream acceptance until a study published more than forty-five years later, in 1981, the year of my birth, reaffirmed that lumpectomy plus radiation was just as effective as radical mastectomy.

Early radiation therapy often took the form of embeddable pellets, but by the mid-twentieth century, radiation beams had become part of treatment. High-intensity beams allowed doctors to treat tumors deeper beneath the skin, where weak beams could not penetrate without lasting exposure that severely damaged skin. Delivering a single tumoricidal dose of radiation also killed too much normal tissue, so doctors spread radiation out over several days or weeks. My great-aunt El and my grandma Meg received their radiation during a time when doctors were still figuring out methods and dosages. Meg received radium crystals for at least one of her two episodes of breast cancer, and her treatment burned a hole in her breast plate that only surfaced several decades later, masquerading, for a terrifying week, as a recurrence. The beams of radiation my great-aunt El received permanently scarred the skin of her chest.

The carnage of war inspired Geoffrey Keynes to pioneer less gruesome treatments involving radiation, but chemotherapy is an even more direct legacy of battle; it's a derivative of the mustard gas that wreaked such havoc during World War I. When studying mustard gas survivors and autopsies of gas victims, scientists on both sides of the Atlantic noticed dried-up bone marrow, damaged lymph nodes, and ulcerated gastrointestinal tracts. This fact—that mustard gas destroys cells that replicate quickly—should have interested cancer scientists, but it lost distinction among WWI's long catalog of recorded horrors. During World War II, however, scientists had another chance to make the leap, after a tragedy near the port of Bari, Italy, in 1943, where German planes attacked a fleet of Allied ships, including one transporting a top-secret shipment unknown to even its own crew: seventy tons of mustard gas. The ship caught fire, pouring its lethal load into the air and water. More than one thousand American soldiers and several thousand Italian civilians died in the catastrophe. Colonel Stewart F. Alexander, who was assigned to investigate the incident, noticed the destroyed lymphatic tissue and bone marrow in the victims and wondered whether mustard gas could do the same thing for someone with leukemia or lymphosarcoma. He sent his results to Yale University researchers Louis

Goodman and Alfred Gilman, who had been studying the effects of the gas in animals. In 1942, they convinced a surgeon to inject one of his lymphoma patients with it. His tumors shrank promisingly but ultimately recurred. After receiving the Bari data, the pair began dosing other lymphoma patients with the drug.

Over the next decade, scientists refined scads of unpronounceable toxic chemicals, from busulfan to triethylenethiophosphoramide, chemicals that kill rapidly dividing cells like cancer. Unfortunately, they massacre other fast-growing tissues, not merely the ones trying to kill the patient, including hair follicles, bone marrow, and the digestive tract, which can cause side effects such as hair loss, nausea, and reduced immune function.

Before the cancer, my aunt El was striking—tall, thin, with determined ice-blue eyes and dark brown locks. After the cancer, she looked different. Her Liz Taylor–thick hair fell out during the chemotherapy. It grew in thin, never regaining its original lushness, and she had to wear it short from then on. The radiation to her entire chest cavity singed the fat around her heart and damaged the surrounding arteries, as well as scarring the skin on her chest. For the rest of her too-short life, she would have stooped shoulders, caved in by the Halsted mastectomies—two years after her original diagnosis, she chose to have her other breast removed prophylactically, an unusual move for the time. Kathy remembers how pleased her mother was when her father purchased a dryer for the family. Thanks to the Halsteds, which had weakened her arms and reduced their mobility, El found it difficult to hang clothes out to dry on the lines. She had lymphedema in both arms, which was painful in the summer, though she never complained.

El responded to the ravages of cancer, in part, with fashion. She'd always been into clothing and enjoyed looking nicely turned out. She bought beautiful custom-made outfits from a couture boutique in Rockford, clothing that looked "very French" to daughter Kathy. But the clothes also served a necessary purpose: due to the physical damage from radiation and mastectomy, they needed to cover her to the collarbone. In particular, El required custom-made swimsuits, and when they went to the pool in the summer other little girls would tease Kathy about her mother's odd concealing swimwear. Because El had grown up with two sisters, Kathy says, "She was very comfortable changing clothes while she would talk to me, or I would run into the bathroom and talk to her while she was in the shower. But this was a lifelong dramatic change. And it was not talked about." Her mother's disfigurement affected Kathy's development during adolescence. "She was the only adult woman I'd seen nude," she says. And because her mother didn't have breasts—not even fat or muscle under the surgical site, "I didn't know if I would have big ones or little ones or what they would look like. There was all this confusion about what they would look like." In contrast, while I was growing up, my mother was very private about her body—I don't think I've ever seen her topless in a changing room. When I asked her about it once, she told me that at the time of her cancer the doctors recommended that women who had mastectomies not appear naked in front of their daughters for fear of frightening them. Years after El's breast cancer, when she was gravely ill with ovarian cancer, El would tell Kathy that the breast cancer hadn't diminished her, that her husband had always made her feel beautiful. He took her dancing. He bought her jewelry. He treated her as just as beautiful and feminine as she'd been before the surgery. Kathy notes that her mother was a product of a time before women's lib, when femininity and attractiveness didn't come from within but rather from how men responded to you. As Kathy puts it, "She always talked about her sense of attractiveness in terms of how my father treated her and responded to her." And he treated her like a woman.

True to the social conventions of the time, El never referred to her cancer as such, but rather as "my surgery." Kathy remembers sitting at the kitchen table with her mother and her mother's friend when she was about twelve—around seven years after El's ordeal. They'd just eaten a meal together. As Kathy tells me, "One of them made a reference to cancer, to my mom having cancer, and I remember it being devastating because no one ever used the word. And I looked at her and said, 'Oh my God, you had cancer.' And she got angry and said, 'Of course I did.' It was a forbidden thing to talk about." Many years later, when Kathy began working at a medical center in Chicago, Illinois, as a psychologist, she would still have to ask the physician's permission to use the word "cancer" with patients she was treating. Some people died without ever hearing the word used to describe their condition—it was thought it might upset patients so much that it would interfere with the regimen of care.

After El's treatment for breast cancer in 1957, my family enjoyed a respite from cancer for the next decade, until a fateful day in 1968. El and Ralph and their younger daughter Lisa were gathered in the kitchen at the end of the day after work. For Lisa, who was ten at the time, the moment exists in horrific, replayable slow motion, each action etched forever in her memory. El opened a letter from my grandma Meg—her sister—and her hand began to shake. Then Lisa's even-keeled, unemotional mother put a hand on the kitchen counter to steady herself and started sobbing. My grandmother Meg had developed breast cancer at only thirty-nine years old. It was the beginning of a very bad decade for the two surviving Muehleisen sisters.

Meg discovered the lump during a family vacation. My grandpa Roy, a professor of horticulture at the University of Arkansas, often attended a summer professional meeting; he'd bring the family along and they'd make a vacation of it, seeing sights and camping along the way. In 1968 they took one such trip to Davis, California, loading the kids—Cris, Gret, Curt, and Mark, who ranged in age from sixteen (Cris) down to eleven (Mark)—into their white Ford Falcon station wagon. It was an era before seatbelts and bucket seats—the kids took turns riding the hump up front, crowded in the backseat, sometimes atop the cooler, or sprawled out in the way back on top of the luggage. Sometimes they stopped at convenience stores for a treat to have with their sandwiches—a soda and a candy bar for each of them. The car didn't have air conditioning or a radio, so they started out each day by singing two hymns and often passed the time by singing other songs. Occasionally my grandpa would depress an imaginary button on the dashboard and tell the kids that would send out the wings and make the miles fly by.

As usual, the family camped their way through Nevada. This trip, they stopped to see the Hoover Dam and celebrated Meg and Roy's wedding anniversary at a park outside Las Vegas, where Roy gave his wife a pearl ring. That night in bed in the tent, my grandpa remembers, Meg turned to him and said, "Roy, feel this. Do you feel a lump here?" He did. It was about the size of an olive.

Because they were past the halfway point of the journey, and because Meg had two cousins, both doctors, who lived a couple of hours away from Davis in Reedley, California, the family headed directly there, where one of Meg's cousins examined her. He told the family to turn around and go back to Fayetteville and even called ahead to Meg's local doctor to arrange immediate treatment. Roy cancelled his meetings.

The couple broke the news to the kids that mom was sick and needed to return to her local doctor immediately. It's unclear whether all the kids learned that "sick" meant "lump in her breast." Cris remembers, "I'm pretty sure it was Dad who told us, and I think he said they didn't tell the boys. I'm not sure when they told the boys." My uncle Mark remembers the car being silent on the way back and that the family stayed overnight at a KOA campground close to a car crash. He remembers seeing "cars mangled up. It was the first time I'd seen a big car wreck like that, and it made a bigger impression on me than anything else on that trip." Dogged by car trouble, the family drove straight home to Fayetteville, sometimes pulling over to sleep by the side of the road.

Forty-eight hours after she'd felt the lump, my grandmother visited her doctor, who told her that she probably had cancer and called for exploratory surgery the following day, a psychologically merciless procedure. In those days, in order to avoid the inconvenience of putting a patient under general anesthesia twice, physicians typically put a woman under, biopsied her tumor, and if the results were malignant, removed the breast without waking her up. My grandmother—like so many other women—fell asleep not knowing whether she'd wake up breastless or with a Band-Aid.

Meg's first mastectomy was brutal. My aunt Cris called it a "butcher job." As my uncle Mark puts it, "I had the sense that they just kind of chainsawed her breast off. When she woke up, this was incredibly traumatic for her. . . . My mom was also a very pretty lady and very proud of her appearance, and this was a very defeminizing experience also, that the medical establishment did this to her without concern for her appearance. . . . This was very, very hard for her." She carried anger about the procedure with her for the rest of her life. A few weeks later, Meg began radiation or perhaps chemotherapy treatments in Fort Smith, a fifty-mile drive down a winding mountain road in a hot un-air-conditioned car under extreme emotional pressure. My grandpa's boss let him take off work to drive her. She required treatments several days a week for several weeks, and they'd return home in time to pick the kids up from school.

Ten years later, Meg lost her other breast to a second bout of cancer. Like her sister El, Meg was beautiful and vivacious, with a cloud of dark hair cut to the current fashion. She'd suffered from insecurity about her physical attractiveness—especially compared to her sister—for much of her life, and she associated her breasts with her femininity and never got over their loss. My grandpa tells me he tried to reassure Meg, explaining to her, "We can go into a restaurant, and you can sit at a table, and some guy across the room can look at you and say, 'What a babe,' and the assessment would be right." He loved her for who she was, not what she looked like. Showcasing his humor, "I told her more than once, I didn't realize until about five years after we were married how good looking you were." He complimented her beauty, explained that he hadn't married her because of her breasts, that he'd rather she lost her breasts than a leg, and that he loved her, but Meg still struggled. Hearing about this decades later from my grandpa, I could barely keep it together; forty years after her mastectomy, my grandmother's same emotions would pour from my mouth, and my husband would offer my grandpa's reassurances in return. Despite the medical advancements that have dramatically changed the nature of mastectomy—the circumstances around our operations could hardly be more different—we had responded to the essential emotion of loss in similar ways. Like Meg and El, I would turn to fashion as a balm for my body fears.

Five years after Meg's first breast cancer diagnosis, in 1973, in faraway Rockford, Illinois, her sister began to feel strange. El's periods became irregular and she spotted between them. She felt full all the time. She gained weight. And her lower intestinal tract seemed angry with her. The doctors couldn't figure out what was wrong—maybe a partial bowel obstruction? Her symptoms worsened, though she and her husband kept them from the younger children, and years passed.

In 1977, Lisa spent one of her last summers at home, between her junior and senior years of high school, working as a seamstress at a bridal shop. Her mom had gotten her the job because she used to buy some of her clothes there, and a driven Italian woman ran the place. Lisa had exactly thirty minutes each day for lunch. On one particular day, she used the time to drive to the hospital with her dad to see El as she woke up from what Lisa believed was a hysterectomy. Lisa remembers, "My dad and I are standing in her hospital room, and they wheeled her in, and her doctors and nurses must have been there. And she was coming out of anesthesia and she said, 'It's bad, isn't it?' " Lisa felt dumbfounded. "[My mother] was visibly horribly upset, my father was upset, and I had to go back to work. I had to go to the old house where the bridal shop was and realize what was happening. 'It's everywhere, isn't it?' It was almost like she knew when she came out of it."

El had stage IV ovarian cancer that had spread all over her pelvic cavity and into her liver. Over the next two years, she endured more radiation and chemotherapy, spending time in and out of the hospital. By 1978—the same year Meg developed breast cancer for the second time—the cancer had spread to El's spine, and she had trouble walking. Because El had taught at the hospital where she received her treatment, her hospital room was a lively place—her many friends among the doctors and staff would come by and chat. El spent about eighteen months slipping in and out of comas, but she never lost her characteristic wry humor. On one occasion, after she'd spent three days in a coma, her eyes popped open and she asked the kids, "Does this make me born again?" Once, when she thought the end was near, she asked Lisa and Kathy to clear out her underwear drawer because she thought that would be too painful for her husband. They packed everything up and tossed it. And then, El lived. She needed new underthings, so Kathy tells me, with a smile in her voice, that El asked them to go to her favorite boutique and buy everything again, but in black and red. And they did it. After waking up from another three-day-long coma she entered on Mother's Day, she told her kids she couldn't have died that day. She said, "I would never do that to you guys—because then how will I have grandchildren if I die on Mother's Day?—because you'll hate Mother's Day." In the years before her death, she'd become interested in Kübler-Ross's five stages of grief (denial, anger, bargaining, depression, acceptance) and would counsel her children about being in this or that stage over her.

In January of 1979, El died.

My grandmother, still recovering from treatment for her second breast cancer, was too distraught to attend her sister's funeral. She'd lost every member of her nuclear family to cancer—her father had succumbed to lymphoma shortly after Trudy's death, while her mother had passed from uterine cancer at age eighty-five.

Cancer wasn't finished with my family. El's dreadful, lingering death kicked off a slew of horrors. In 1981, the same year my grandparents welcomed their first grandchild—me—into the world, Roy developed colon cancer, initially misdiagnosed as terminal liver cancer, which he survived. In 1985, my grandmother would develop stage III ovarian cancer—a huge psychological blow for someone who had already lived through cancer twice—and a horrific physical ordeal that she survived against all odds. The mortality rate for ovarian cancer hasn't budged much since 1940—even today, only 44 percent of patients diagnosed with ovarian cancer survive for five years or more.

In between my grandpa's colon cancer and my grandma's ovarian cancer, in 1983, my parents finally took a honeymoon. They'd been married for five years and lived in Dallas, where my dad worked as a tax lawyer. They stashed their eighteen-month-old (me) with my father's parents and headed to Hawaii for a few days. In an eerie echo of my grandmother's breast cancer diagnosis, for a long time my father claimed that my mother found the lump on vacation, his memory muddled by time and emotion, this worry conflated with the actual bad news some months later. In fact, my mother had had a routine physical that included a mammogram right before the trip. While she and my father vacationed, the doctor's office called to explain that the scan had been clean.

Four months later, my mother found a lump on the front of her right breast while showering and immediately scheduled a doctor's appointment. Since her recent mammogram had been fine, her doctor suggested they adopt a watch-and-wait approach for a few months and prescribed some vitamins. A few weeks later, with the lump unchanged, she had an uninformative needle biopsy. Then came the surgical biopsy. Thanks to the women's health movement of the 1970s, my mother did not fall asleep uncertain of whether she'd wake up sans her right one. They woke her up, delivered the unfortunate results, and performed the mastectomy the following day. The verdict—rendered a few weeks short of my mother's thirty-first birthday—was stage II infiltrating lobular cancer of the right breast. Her tumor had grown from nothing—as indicated by her recent mammogram—to a monstrous four centimeters in only four months. The simple mastectomy removed her entire breast but not the underlying chest muscles, as a now-outdated Halsted would have. Luckily, her biopsied lymph nodes showed no signs of involvement. And thanks to the advancement of medicine, she had her cancer tested for estrogen receptors as well.

In the 1890s, Scottish surgeon George Beatson had experimented with removing the ovaries, first of sheep and cattle, and then of breast cancer patients, and discovered that ovarian removal could shrink breast tumors. In the 1920s, scientists identified the hormone estrogen, but the big breakthrough wouldn't come until the late 1950s. Canadian surgeon Charles Huggins discovered that certain hormones could perpetuate the growth of prostate cancer—he would later win a Nobel Prize for his research—and he wondered whether a similar mechanism might be at work in breast cancer. He experimented with the removal of the ovaries and the adrenal glands, which also secrete sex hormones, including estrogen. He found that the removal of these organs beat back cancer—at least temporarily—in 30 to 40 percent of advanced breast cancer patients. At the time, there was no way to tell who was in the group that would benefit; so he urged Chicago chemist Elwood Jensen to look into it, and Jensen did. As it turns out, some breast cancers contain receptors that are stimulated by estrogen, which promotes the cancer's growth, and some do not. Over time, scientists would discover that certain breast cancers may also express a fondness for progesterone or contain a lot of HER2/neu, a protein that occurs on the surface of some tumors and can promote cancer growth. Soon, therapies targeting these sorts of cancers would come on the market, adding hormone therapy and drugs that target HER2/neu pathways to the modern arsenal of cancer-fighting agents. Breast cancer related to a BRCA1 mutation—as my mother's cancer was, though of course we did not know it at the time—is more likely to strike younger women and is more likely than non-BRCA breast cancer to be negative for all three receptors.

At the time of my mother's cancer, only the estrogen receptor (ER) test was available, and hers came up negative. Many women would consider ER-negative cancer unlucky, because ER-positive cancer provides physicians with an extra weapon against the invader: deprive the beast of the hormones it feeds on and watch it die. My mother chose to view her ER-negative cancer positively; it meant she could take hormones for menopause far down the road without worries about feeding some itinerant cancer cell. She received six chemotherapy treatments and thirty radiation treatments. A few months after her diagnosis, she found a lump in her other breast. The doctors believed the lump to be benign, but she decided to take that breast off prophylactically.

Though it wasn't known then, current research shows that most women with cancer in one breast do not benefit from having the remaining breast removed preventively, a procedure known as contralateral prophylactic mastectomy. For women with sporadic (non-hereditary) cancer, several studies show that the risk of developing a second breast cancer in the healthy breast is relatively low—around 3 to 9 percent. In addition, having the second breast removed does not significantly improve long-term survival rates—a 2013 analysis found that it improves life expectancy by a maximum of six months. At the same time, the operation reduces quality of life. For cancer patients with a BRCA mutation, however, the risk of contralateral cancer is much higher—about 47 percent over twenty-five years after the initial diagnosis. For these reasons, the National Comprehensive Cancer Network, for example, discourages contralateral mastectomy unless a patient is at high risk, as my mother was.

My mother, of course, didn't have the benefit of this research, so she made the best choice for her, according to the knowledge she had at the time.

A lifelong feminist, she hadn't changed her name after she married my father, but she did so now. As she wrote to me decades later, "I was so worried about you. I was worried that I wouldn't make it," and so she changed her name, "thinking that if I died you would not remember me and having the family name would help with that." It was sad to lose her breasts. "My breasts were part of who I was as a person; it was part of my sexuality and it was part of my having experienced motherhood. It was a huge loss." The blow to her sexuality, she says, might explain her strong memories of two events after the surgery. One of the wives at the law firm gave her a pretty nightgown, a slippery lime-green thing with short sleeves. "I thought it was nice to have something so feminine and pretty," she says. She'd never had an engagement ring, and after the surgery, my dad presented her with a diamond ring, a romantic gesture that signaled to her that he cherished her for who she was. And his attitude was loving—he never made her feel self-conscious about her body, or as if she was less of a woman. "But it's a sexual loss. There's no doubt about it," she tells me. She would look at herself in the mirror and feel like she looked like a kid again without her breasts. Her missing breasts presented a challenge she had to overcome. "When they're gone, you have to readjust how you think about yourself," she says. "On the other hand, because sexuality is a total body thing, having a piece of it missing isn't going to erase the sexuality, it's just going to alter it. It was sad, but at that point too, I was glad to be alive. What made me more sad at that time was the thought that I might not see you grow up."

The missing pieces of her weren't the worst part of the whole thing, she says. Having cancer meant shuttling between terror and normality. "There were nights during that period where I'd wake up in the middle of the night and just weep and think, 'Oh, I'm not going to see Lizzie grow up and what's going to happen to me and I'm going to miss Dick [her husband and my father] so much. You have those moments—there's just no denying that—and during the day you just get up and have breakfast and read to your daughter and go to the park."

The cancer meant something major to the rest of the family too. My mother's diagnosis was extraordinarily upsetting for her parents, for her mother in particular. No one quite remembers how the news spread to Lisa and Kathy—my grandmother or my aunt or my mother wrote letters or called the Illinois branch of the family—but my mother's diagnosis terrified her relatives. Kathy hadn't waited long after her mother's death to seek out a then-controversial prophylactic mastectomy, a move my mother had initially thought was "a little extreme." As for Lisa and Cris, of course they suspected the disease might be hereditary—Trudy, El, and Meg's cancers certainly suggested that as a possibility—but after my mother fell ill, the Muehleisen curse felt undeniable, biblical, and impossible to avoid. As Lisa puts it, "When I found out about Gretchen I thought, oh, it's going into the second generation. The prophecy is being fulfilled. This horrible thing is coming down into this next generation."

Though none of us would know cancer, we would know the curse of fear. And with a scalpel, we would choose to cut it out.

3 | Gene Hunters

The first or second time I came home from college, I spent many late nights in my mother's home office, using her computer and the house's only Internet connection to stay in contact with my new friends, chafing, undoubtedly, at the bonds of parental supervision that I'd been free of for the last few months. But if my parents could watch me—an adult—I reasoned, then surely I could watch them. Perhaps this explains why I went snooping. My fingers walked through the file drawer in the desk until they settled on the cancer folder. I just wanted a peek, I told myself. Just a peek.

I paged through a bunch of things that didn't interest me—medical papers detailing my mother's treatments and family trees—until I lit on a piece of personal correspondence she'd written to one of her friends in Texas. We'd lived in Dallas until I was five, and my mother had run in the very first Susan G. Komen Race for the Cure there in 1983, only months after finishing breast cancer treatment. As she wrote to me years later, since then "I have been able to do all the things I love—hug my daughter, play tennis, garden, shake my fist at injustice—all sorts of things." We'd moved to Washington, DC, in 1986, though of course my mom still had friends back in Texas. In

1991, when I was ten, she'd written a letter to one, a politician who briefly sat on one of the Susan G. Komen boards. My mother begged the woman to vote to fund research into the new so-called breast cancer genes, not on behalf of herself but on behalf of her ten-year old daughter—me. Reading this letter nine or ten years later scared me. I'd stumbled onto something unexpectedly personal and intimate, proof of my mother's love and of her fear. She must have been so scared for me. I cried over the note but put it carefully back in the file and said nothing.

While my mother wondered what might be in our genes, an extraordinary scientist and human being had already made possible this search for the specific BRCA mutations. So let us sing the praises of Mary-Claire King, scientist and citizen, whose humanity and goodwill toward others is well known, whose deep engagement in the world has reunited families and revolutionized genetics, a remarkable woman of uncommon scientific stamina and ambition.

I've never spoken with Mary-Claire King, but as you might have guessed, I'm a bit in love with her from afar, sort of in the same way I'm in love with Dorothy Parker and Susan B. Anthony, a distant lady crush sparked by reading her words and about her work. Mary-Claire King is the Eleanor Roosevelt of science—an advocate for social justice and human rights, a prolific contributor to her chosen field of genetics, and an inspiring feminist. Mary-Claire King once turned down a shot at the directorship of the National Institutes of Health (NIH) because she thought it would take her away from her addiction, the daily practice of science! Mary-Claire King puts a premium on hiring women and ethnic minorities and supports the new mothers who work for her by providing pumping space in the lab! Mary-Claire King is fostering a rare scientific collaboration between Israeli and Palestinian scientists seeking a gene for inherited deafness! Mary-Claire King once single-handedly wrestled a newborn rabbit away from a gorilla! OK. So maybe that last one isn't true. Although her résumé is full of accolades and impressive finds, she's best known for three scientific feats.

She performed the first one in the 1960s and 1970s while still in graduate school at Berkeley, where she initially studied mathematics. While there, she met anthropological geneticist Allan C. Wilson, a renowned scientist who would later become famous for his work on mitochondrial Eve—the hypothesis that all humanity descended from a single woman, "Eve," who lived more than a hundred thousand years ago. Mitochondria live inside all of our cells and are responsible for creating the energy that each cell uses. They also have their own DNA that is distinct from the DNA in the cell nucleus. And we inherit our mitochondrial DNA from our mothers only—sperm do not contain mitochondria, but eggs do. In this way, matrilineal lines may be traced by examining this unique form of DNA, a fact King would later put to good use. Wilson's theoretical and experimental discovery of mitochondrial Eve earned him worldwide recognition in the late 1980s. But back in the late 1960s and early 1970s, he convinced the young King to enter the bourgeoning field of genetics, and he advised her dissertation, which tackled another aspect of evolutionary genetics, namely, the similarity of human and chimpanzee DNA. Her doctoral research proved that the two species are 99 percent genetically identical, leading to the conclusion that it is not merely genetics that shapes our biological processes but a complicated system of timing, with cells turning proteins on and off at particular moments. King and Wilson published their results in *Science* in 1975 and earned worldwide recognition for the discovery.

At the same time, working in the Wilson lab taught King many attitudes that would shape her practice of science for decades to come. Wilson taught her to be "very methodical and very self-critical" but to trust absolutely in the data unless someone else had better data that proved her wrong; he also taught her to never let anyone intimidate her. As she told an interviewer years later, "You did not grow up in the Wilson lab thinking that you were right! We were extremely critical of ourselves and of each other. Persistence was also part of the culture of the Wilson lab and it is part of me." Working with Wilson taught her to become comfortable living with years of uncertainty—to me it sounds a bit like living with a BRCA mutation.

At the same time, King was also deeply politically engaged as an activist agitating against the Vietnam War. As she told an interviewer more than four decades later, "I don't indulge in nostalgia about that period because it was a terrible time for our country. It's very American: When you see something wrong, you try to fix it." The best thing the activists King worked with did, she said, was put on their professional clothes the day after the United States invaded Cambodia, "not clothes that any of us had worn since coming to Berkeley," she noted. They marched down to synagogues and churches. "By the end of Sunday we had 30,000 letters opposing the action. It made it longer to get a dissertation done, but it was an interesting, intense time," she said. At one point, she even dropped out of school briefly to work on an environmental project with consumer activist and later Green Party presidential candidate Ralph Nader. But in the end, research reeled her back in, and as it turned out, her political sensibility was eminently compatible with her scientific ambition.

King once told an interviewer, "I've learned not to question the motives of bastards. They just do what they do, and you try to stop it." And as it turns out, she can stop bastards—with science. The second major accomplishment in her lifetime achievement hat trick combines genetics with justice. After finishing her dissertation, King moved to Chile to teach genetics, a stint cut short when a military coup deposed the government of President Salvador Allende and killed him. Although King and her then husband, a zoologist, moved back to the States, the episode left her with an understanding of the political situation in parts of South America. Her genetic expertise and political sympathies soon converged with the needs of a civilian activist group in Argentina, the Abuelas de Plaza de Mayo, a group of grandmothers seeking their lost grandchildren.

During the Dirty War of the 1970s, Argentina's government—a dictatorship supported by the United States—"disappeared" thousands of so-called dissidents, including whole families with children. In some cases, the victims were pregnant women who were tortured and killed and who may have given birth in prison. Many of these taken or prison-born children—who would be about the same age as King's own daughter—had been sold to or illegally adopted by junta supporters. In 1977 the Abuelas de Plaza de Mayo, the grandmothers of these children, banded together to find them; the group drew its name from the plaza where they protested in front of the junta's headquarters in Buenos Aires. They began gathering data about children who might have been illegally adopted, and they searched for a geneticist, settling finally on Mary-Claire King. She was an ideal choice both because she had a connection to the region and because she had done work on human DNA. They asked her to produce a test that could prove a genetic connection between these kids and their grandparents. First, King developed a test using blood proteins to prove grandparentage, and later she worked on mitochondrial DNA testing, which could prove matrilineal descent. In 1984, the Argentinian Supreme Court ruled that a child identified by King's test had to be returned to relatives, setting precedent for later reunions. In early 2014, the Abuelas found their 110th missing grandchild, though her grandmother had already died, according to the Abuelas' press release. The identity of many of these children was confirmed through DNA testing.

King went on to develop a process to extract mitochondrial DNA from teeth, a useful tool for identifying bodies—many aged human remains don't have enough skin or hair remaining to make DNA extraction possible, and DNA from bones is easily damaged. Watch out, murderous dictator bastards: her lab has helped UN war crimes tribunals identify remains of victims of violence in Cambodia, Guatemala, El Salvador, Rwanda, Ethiopia, and Bosnia. Though her lab operated for a time as the forensic lab for these tribunals, now that the science of forensic genetics has improved across the board, the King lab is mainly consulted for special cases. It has helped identify the remains of the Romanovs, the last czars of Russia, as well as MIA US soldiers from WWII, Korea, and Vietnam.

When King undertook this forensic science project of social justice she was already deep into her research on genetics and cancer risk. The link between certain families and cancer had been known for more than a century. In 1866 Parisian surgeon Paul Broca—known primarily for his discovery of one of the brain's language centers—had published a pedigree of one such family tree, going back four generations; the family was his wife's. Of the twenty-six relatives she had over age thirty, fifteen of them—all women save one—had developed or died of cancer, a staggering 74 percent of the women and 14 percent of the men. In contrast, the rate of cancer in the general population at the time was about 3 percent. These numbers convinced Broca that cancer could be inherited, although the mechanisms of inheritance would remain opaque for a long time.

An early study on hereditary breast cancer published in Denmark eighty years later would establish the process for studying cancer families. The study examined two hundred families and used the national cancer registry to help generate pedigrees—researchers tracked down relatives of those who suffered cancer. Finding patients, questioning them about their own cancer, seeking out distant relatives with disease, and finding their pathology or autopsy reports to verify the presence of cancer was time-consuming but necessary work. The study showed that women with a first-degree relative—parent, sibling, child—with breast cancer were at a higher risk of developing the disease, a conclusion that a similar 1948 British study confirmed.

Yet hereditary breast cancer proved difficult to study. For starters, many scientists still doubted that cancer had any hereditary component. After all, families share much more than just DNA. A family of smokers might all develop lung cancer, but that wouldn't necessarily make their disease hereditary—it could be that the parents passed on their pack-a-day habit to their kids. Or maybe they all lived on top of a landfill full of nuclear waste. Environmental and cultural causes made looking for hereditary disease tricky. So did the fact that most cancer is sporadic. Today, about one in three women and one in two men will develop some sort of cancer in their lifetimes, according to the American Cancer Society (ACS), so it makes sense that most families will have more than one case of cancer in them. But that doesn't mean that these cancers are necessarily inherited—even cancer families have some cases of sporadic cancer. I have inherited breast and ovarian cancer syndrome in my family, but not all of the cancer in my family is linked to our BRCA mutation. My mother's thyroid cancer, her father's colon cancer, and on my dad's side of the family, relatives with prostate, stomach, ovarian, and lung cancer all occurred independent of the Muehleisen curse. Those cancers are just the roll of the dice we all make by living. Inherited breast cancer was particularly confusing to study. Because breast cancer is one of the most common cancers in women, there are many families with multiple cases. And finally, as Ilana Löwy, senior researcher at the French National Institute of Health and Medical Research, reports, "Researchers initially confused two distinct phenomena—a relatively moderate (two- to threefold) increase of incidence of breast cancer in daughters or sisters of women diagnosed with this disease, and the existence of 'breast cancer families' with an exceptionally high frequency of this pathology." For all these reasons—the commonness of sporadic cancer in general and the frequency of breast cancer in particular, the possibility that familial cancer might be caused by lifestyle or environmental factors, and the presence of two similar cancer syndromes—the whole issue was hard to study.

Therefore, scientists needed large families because large sample sizes give more accurate predictions than small ones. If you do a poll at my house about the color of human hair, you'll conclude that humans have dark hair, because that's what my husband and I have. Poll my entire neighborhood, and you'll have a much better idea. Hereditary cancer is relatively rare—searching for it meant looking for the proverbial needle in a haystack, or more accurately, the proverbial needle in a field of haystacks. Looking at large families gave researchers a better shot at examining the syndrome. If a couple has eight kids and seven of them end up with breast cancer, that family is probably better to study than a family with four kids and two cases of breast cancer. Sure, it's possible that both families simply suffer from an unlucky bout of sporadic cancer, but if you're looking for a familial link, it makes sense to study the ones that seem stronger first. The larger the family, the better, which meant painstaking research to trace familial cancer in large families over many generations, even before the human genome had been sequenced.

Omaha physician Henry Lynch began collecting such pedigrees in the 1960s for colon and endometrial cancer. Colleagues in the field would dismiss and ignore his work for decades, until science's growing knowledge of genetic sequencing improved and other theories of cancer transmission—such as the idea that it was caused by viruses—fell by the wayside. Lynch insisted that some cancers were hereditary and argued for preventive surgery as a treatment; eventually he became interested in breast cancer. By that time, in the 1970s, geneticists had already figured out that women from breast cancer families often developed the disease at unusually young ages and frequently suffered cancer in both breasts.

Mary-Claire King joined the search for hereditary breast cancer in the mid-1970s and would spend most of the next two decades exercising the persistence she'd acquired in Wilson's lab. She had a personal connection to cancer. Her best friend since the age of four had died from kidney cancer, when the girls were both fifteen, after a long painful battle with the disease, an unjust, too-young death that made a deep impression on King. Although she doesn't have breast cancer in her own family, in 2004 she told the *Chicago Tribune*, "I was interested in it because it was a disease of women. I was interested in the familial form because it seemed to strike young women disproportionately." And as she began her research, "it's been very much a story about people I've come to care about enormously. The participants in this kind of work are very sophisticated about there being no quick answers." An unabashed feminist, King speculates that the public feels so keenly about breast cancer because "almost uniquely among cancers, breast cancer is a cancer of women who are successful. The longer the interval between one's first menstruation and first birth, the higher the subsequent breast cancer risk. It's a cancer of women who are educated, who have children later in life. Whereas many illnesses are the consequence of poverty, breast cancer is closely tied to success."

While working on a project at the University of California, San Francisco, searching for risk factors linked to breast cancer, King had a eureka moment. She had pored over thousands of questionnaires filled in by breast cancer patients, trying to categorize all the different risk factors, when "I was struck by something that others had seen before and that seemed so obvious I couldn't think of anything else," she told author Michael Waldholz. Many of the women stricken also had a close relative with the disease. "Given my background [in genetics], it's understandable that I homed in on inheritance."

In 1975 she landed a job at the University of California at Berkeley, thanks to the presence of the only woman on the hiring committee and to the new policies of affirmative action. "After I had accepted the job, the division head said to me, 'I just want you to know that you are only here because of all these new regulations, and we are really scraping the bottom of the barrel in hiring you.' And I said, 'We'll see how long you feel that way!'" she told *PLOS Genetics*. Mostly, as she has told numerous interviewers, being a woman in a male-dominated profession helped her because other researchers didn't see her as a threat. On top of this she was working in the unfashionable field of inherited cancer, which ran contrary to the current hot theory of viruses. As she told the *Washington Post* years later, "In retrospect, there was something liberating about being of no interest to the leaders of my field. If one is ignored, there are no expectations to meet. There was great freedom in being invisible."

King's goal was to get enough good data about cancer families to build a testable mathematical model. So she found her way onto a National Cancer Institute study investigating whether birth control changed a woman's risk for breast, ovarian, or uterine cancer and persuaded them to add a couple of questions about family history. "It wasn't too long before the family history questions threatened to overwhelm the project! The interviews were fabulous, and we ended up with 1,500 pedigrees based on reports of cases and another 1,500 based on reports of controls," she told an interviewer years later.

King worked the case the old-fashioned way: with lots and lots of tedious work. Though we now know that humans have around twenty thousand genes, at the time King began her research we didn't know what the upper ceiling was. Only a few hundred genes had been located on chromosomes, and only a fraction of those had been sequenced at all. So in order to prove that breast cancer risk could be genetic, King had to sequence an unknown gene on an unknown chromosome that was maybe or maybe not shared by members of large cancer families. First things first, though—she needed a bunch of huge cancer families to begin her research. Suddenly, Henry Lynch's work in Omaha became a treasure mine.

Lynch had the pedigrees but not the technical know-how to look inside their DNA, while King had the know-how but needed families; so they collaborated for a while. Later, in 1987, First Lady Nancy Reagan would inadvertently help King find families too. A local television reporter interviewed King for a spot on the birthday of the National Cancer Institute, and she explained her project and the need for cancer families. The next morning, Nancy Reagan announced her breast cancer diagnosis, and news media scrambled to find spots on the topic. King's interview was rebroadcast across the country, and letters from families came pouring in.

King used some of the sequencing techniques she'd honed during her dissertation, namely the use of polymorphic gene markers to investigate the families. Such markers help scientists locate genes by providing a landmark of sorts. Let's say George is in New York City and I want to find him. Well, New York City is a huge place full of buildings and people—looking in every nook and cranny, assuming he stays still, would take several lifetimes. But if I know that George usually hangs out within five blocks of the Empire State Building, that narrows things down considerably. It'll still take a long time to find him, but it'll be much faster than ransacking the entire region. A polymorphic marker operates like the Empire State Building—it helps scientists know they're at the right part of the genome.

Physical traits can serve as gene markers. If everyone in my family who has asthma also has attached earlobes, then perhaps the genes that code for these two traits—asthma and earlobes—are inherited together. And if they are inherited together, perhaps it's because they are located right next to each other on a strand of DNA. Scientists figured out how to find markers not linked to physical traits and used those to help study heredity.

So King spent the next decade and a half searching for gene markers present in cancer patients from families with lots of cancer. If her lab could identify a good marker, then she'd be able to tell which gene might hold the key to familial breast cancer risk. Eventually, she and her team assembled nearly two dozen cancer families from the United States, Puerto Rico, Canada, the United Kingdom, and Colombia. They unearthed hospital reports and death certificates to confirm the presence of cancer in some people and drew blood from the 329 surviving family members so they could extract DNA. Her team was looking at 173 different markers in the twenty-three families, and as she told an interviewer, "of course everything was done by hand because none of the analysis was computer-based. We literally had the pedigrees rolled out across lab benches and floors. Then Beth Newman [a colleague] had the idea of arranging the pedigrees in the hallway in order by average age of breast cancer diagnosis in the family." After all, they knew that women who developed breast cancer young were more likely to live in families with many cases of cancer. Focusing on these women, they hit pay dirt. At a genetics conference in October 1990, King announced the shocking findings: one part of chromosome 17 was altered in women from breast cancer families who had the disease. And as many as one in two hundred women carried such a mutation, making it one of the most common inherited syndromes. King still felt a bit unsure of herself. At the conference, "I presented our data as an interesting story. I knew it was statistically robust, but I was concerned it might be some elaborate fluke." Several important research groups—including those of French scientist Gilbert Lenoir and Utah-based geneticist Mark Skolnick, who would put his own stamp on the breast cancer genes soon, were in the audience and asked King for her marker sequences, which she sent out. Two months later, the King lab published its results in *Science*. A few months after that, in early 1991, Gilbert Lenoir—then a collaborator of Henry Lynch—presented a talk on hereditary breast cancer at a conference in England. "He presented what I interpreted as a summary of my talk," King told an interviewer, "with extremely familiar results, assuming that he would go on to describe his results next. But he stopped. I asked what were his results, and he said, 'Those are mine!' His results were virtually identical to ours! Same markers, same age effect, same lovely fit to the same model, exactly our results but on unrelated families." Lenoir had confirmed her results and also improved slightly on King's findings, showing that women with an alteration on this part of chromosome 17 had a high risk of ovarian cancer as well as breast cancer.

Many years after snooping in my mother's desk, I confessed and asked if I could see the letter. She dug it out, along with two other letters of note. We discovered that in 1991, the year King's announcement was verified, my mother read about it in the paper and immediately sent letters to the King and Lynch labs offering the family DNA for study. She remembers answering questionnaires for both.

King's news fired off the starting pistol in a heated race to find the gene, dubbed BRCA1. Although, of course, "race" isn't the word King would use. As she told Waldholz, using that word "is just an awful way to describe what we're doing, especially when you think about the women who are depending on us. I don't mind this being called a race, if you mean a race against the disease. Every time we hear of another woman who died of breast cancer, we take it personally. We should have the gene by now." In the twenty years it took to find the gene, King pointed out, more than one million women died of breast cancer. Later, in her 2014 piece recounting her search for the BRCA1 gene, she explained, "Until there are no more breast or ovarian cancers among women with BRCA1 or BRCA2 mutations, the real race is not over."

King's 1990 results proved that some breast cancer was genetically linked—but there was still plenty of work to do. DNA has that familiar twisted-ladder structure, in which each rung represents a chemically joined base pair. Guanine binds with cytosine and adenine binds with thymine to make our DNA strands. The twisted ladder, which can be unzipped to make proteins that regulate cell function, is itself wrapped around proteins called histones, coiled up into larger structures called chromosomes. Humans have twenty-three pairs of chromosomes—one set from each parent—wound up inside the nucleus of almost every cell in the body. King's results were astonishing because she'd narrowed down the gene's location from three billion base pairs—the size of the whole human genome—to a particular region of DNA that was twenty million base pairs long.

But twenty million base pairs is still pretty big. As geneticist Francis Collins put it to Waldholz,

> If you consider human DNA as being the size of Earth, King had just placed the gene somewhere in Texas. It was no small accomplishment because few people even believed it had existed at all. Yet, now the job ahead was to pinpoint the gene, to first map it to a particular county in the state, then to a town, then a street, then a house on the street, and finally, the exact room in the house. It was going to be a very difficult assignment.

The prize—to be the first to clone the BRCA1 gene, earning fame and funding, and maybe saving some lives—enticed the world's best genetic laboratories into the arena. According to King, more than one hundred researchers from at least twelve labs spent the next four years searching for BRCA1, a particularly extraordinary effort, given that the Human Genome Project had just begun, so the hunt started nearly from scratch.

Mark Skolnick's lab in Utah was already in the fray. Skolnick had begun his career looking at population genetics. In grad school he worked on a project tracing the family trees of the residents of Parma, Italy, to study the genetic shifts that occurred over time in a populace that had been mostly isolated. Thanks to his skill with computers and his previous background in demographics, his adviser tasked him with organizing centuries of family history. After the project ended in 1973, a professor at the University of Utah contacted him in hopes that he could apply the same know-how to help find cancer in the Mormon families of Utah.

The Mormon family trees represented a huge advantage in the race to identify the breast cancer gene for several reasons. For starters, the Mormon Church encourages large families, and large families are good for cancer genetics research. The religion also believes that people can be baptized and receive the gospel after death. As the church website puts it, "Discovering that you're related to a renaissance nobleman could be fun. It could also mean giving him and his family an opportunity to receive the gospel of Jesus Christ." In practice, this means that tracing family trees is a holy mission, and the church has assembled an incredible number of family pedigrees—over one billion people tracked down by Mormon sleuths—that would prove to be a treasure trove for researchers like Skolnick.

In addition to the Mormon genealogical records, Utah had an extensive cancer registry that started compiling current cases in 1966 and had records on tumors that dated back to 1952. The potential juxtaposition of the two databases intrigued him—combined, they could be a powerful tool. The university didn't have funds for such a large project, so the young Skolnick applied for funding from NIH, despite the fact that prevailing wisdom didn't favor a link between cancer and genetics. In the meantime, he made a curious discovery in the genealogical records: the thirty-three cases of rare male breast cancer he found were all related to one another through relatives near and far. He also studied lip cancer, learning that it clustered in families too. He decided to investigate breast and colon cancer. After receiving the grant, his team spent the next five years on data entry, linking two hundred thousand family pedigrees that included records on about 1.6 million individuals from the church's library with thousands of cancer cases from the state registry to form a database. As Skolnick identified possible cancer families, his lab collected blood and tissue samples from some twenty thousand people over a decade.

In the meantime, because cancer risk is incredibly complicated, Skolnick went after a simpler hereditary disease in hopes of developing a method that could be used to attack the problem of finding a cancer gene. He focused on hemochromatosis, a rare disorder that prevents an individual's body from processing iron properly, leading to buildups that can cause cirrhosis of the liver, diabetes, and heart damage. Researchers already knew it was hereditary but weren't sure whether it was a dominant trait, requiring one copy of the gene, or a recessive trait that required two copies of the gene. Building on French research that found a blood protein associated with hemochromatosis, Skolnick figured out that the gene coding for that protein must be a marker for the hemochromatosis gene, which helped locate the gene responsible, proving that it was recessive and leading to treatments such as blood transfusions for people who had not yet developed the full-blown disease.

And then, at a ski retreat in 1978, Skolnick, MIT biologist David Botstein, and geneticist Ronald Davis of Stanford University had a brainstorming session that would change the landscape of genetic research. Essentially, they figured out how to make markers—like the DNA coding for the blood protein had been for hemochromatosis—synthetically using enzymes. Different enzymes snip DNA strands at specific places, say between the A and the G (adenine and guanine, two bases that make up the rungs of the DNA ladder, along with thymine and cytosine) in the sequence TCTCAG. But human beings have a lot of variation, and some of us have longer genes than others. The enzyme still snips our DNA strands at the same place, but in me, the resulting fragment might be longer than yours, with an extra string of letters in there. Those letters, known as restriction fragment length polymorphisms, or RFLPs, can serve as gene markers and are passed down through generations. If I have my mother's RFLP, perhaps I also have her breast cancer gene. If everyone with cancer in my family has that RFLP, then perhaps it could be the genetic Empire State Building, guiding researchers ever closer to the location of the BRCA1 gene. The discovery of RFLPs revolutionized genetic research.

Months after Mary-Claire King announced that she'd narrowed down the location of the gene, Mark Skolnick joined forces with businessman Peter Meldrum to form Myriad Genetics "specifically with the goal of isolating the breast cancer susceptibility gene," as Skolnick told the DNA Learning Center. Skolnick needed funding in order to compete with other gene hunters. As he put it in a deposition for the later Supreme Court case around the BRCA1 gene, "I was also keenly aware that NIH had awarded [renowned geneticist] Francis Collins a massive genome center grant which would allow him to pursue cloning this gene, and that my group would most likely not be given adequate funds to compete. This in fact turned out to be true. My collaborators and I submitted a small grant proposal to pursue BRCA1, but we were turned down. We were told that we didn't have the family material to be competitive, when in fact it was common knowledge that we had spent years collecting the most extraordinary breast cancer families in the world." If Skolnick couldn't get the money through grants, then he'd get it through private funding. By August 1992, Myriad had received a $4 million investment from a pharmaceutical company and added Dr. Walter Gilbert, a Nobel laureate in biology, to its team. By March 1993 they had $8.8 million more in investments, according to Skolnick's deposition.

With considerable funding at his disposal, Skolnick was hell bent on success. He stated in his deposition, "We were acutely aware that if we were to fail to find BRCA1 we would have had great difficulty in surviving as a company and that our jobs would be lost." By the end of 1994, after many long hours in the lab, they had identified and cloned the BRCA1 gene, beating out some of the world's best laboratories in the process, including King's.

"People ask me how I felt," Skolnick told Waldholz. "You know, we didn't pop champagne, we didn't go out to dinner. We just felt this tremendous relief. After all these years, I could finally take off my racing shoes." Skolnick and King had had a brief collaboration around the breast cancer gene a decade earlier but had gone through an ugly split. A 1994 *New York Times* article about the discovery mentioned that King, "whose personal less-than-tender feelings about Dr. Skolnick are well known to her colleagues, nonetheless described the discovery as 'beautiful' and 'lovely' and deserving of all the praise it might win." Later, she would tell the *Lancet*, "While the race was in progress I certainly thought that I would be disappointed if we didn't win. But actually, much to my surprise, I was not."

With its commercial motives, Myriad went on to patent the BRCA1 gene in a move that caused immediate controversy. But, as an October 29, 1994, article put it, "the gene's co-discoverers, the University of Utah and Myriad Genetics Inc. of Salt Lake City, left their Government collaborators off the patent for BRCA1. That meant the National Institutes of Health would lose control over diagnostic tests or cancer therapies developed from BRCA1 and their prices." Myriad president Peter Meldrum told the Associated Press that the company didn't include NIH on the patent because it hadn't made a significant contribution to the discovery. The lead scientist for NIH begged to disagree. And while Myriad said it had spent $14 million discovering the gene, NIH contended it had contributed $4.6 million.

Though the patent disputes over BRCA were just beginning, the search for hereditary breast cancer was not yet over, because BRCA1 did not account for all the many cancer families scientists across the world had been digging into. This problem fascinated Dr. Mike Stratton, a scientist with the Institute of Cancer Research and the Wellcome Trust's Sanger Centre in the UK. Unlike King and Skolnick, Stratton had a strong medical background, beginning his career as a histopathologist, someone who looks through the microscope at tissues—in Stratton's case, cancer cells—to decide whether they are healthy or harmful. "I knew the DNA was abnormal, but I couldn't see what the abnormalities were," Stratton tells me. "So it seemed to me a fascinating and all-engrossing question to find out what those abnormalities were." At age thirty he left medicine, earned a degree in the molecular biology of cancer, and started developing the Breast Cancer Genetics Programme at the Institute of Cancer Research in London.

After the discovery of the BRCA1 gene, he began collecting cancer families without BRCA1 mutations in hopes of discovering another breast cancer gene. At first, he gathered families in southwest England, people who had kids in the 1980s and 1990s, but it proved a nonstarter—the family groups were too small, and people often lost contact with the relatives who were so essential to track in a study like this. He needed bigger families, so like Skolnick, he worked the religious angle, deciding to focus on the nearby Catholic country of Ireland. He sent letters to all the oncologists in the country, and through these connections he located one particularly large, particularly cancer-prone family. Over a year and a half, his team tracked down several hundred members of this family stretching over four generations; the family had about thirty cases of breast cancer not linked to BRCA1. And then they learned about a case of male breast cancer, something that is rare even among BRCA1 families. Stratton's team added their discovery—male breast cancer in a non-BRCA1 cancer family—to the evidence that there had to be another breast cancer gene. As Stratton puts it, "This family became the centerpiece of our research for BRCA2. We now had evidence, strong evidence, that there was another gene."

Stratton's lab in the UK formed a collaboration with Skolnick's lab in Utah to find the general location of the BRCA2 gene, which was now a relatively fast process because maps of markers throughout the genome had improved since King and Skolnick had begun their work more than a decade earlier. Stratton and Skolnick located BRCA2 on the long arm of chromosome 13 in record time—about a year. The next step was to locate the BRCA2 gene itself inside the arm and to clone it.

But Stratton worked for a nonprofit institute, the Sanger Centre, and Skolnick worked for a for-profit company, so they parted ways. "We'd had a cordial collaboration," Stratton says, "but if we continued that collaboration to the identification of the gene itself, that would be with Myriad, and we were not comfortable with the patenting and monopolizing policy of Myriad." The race was on again, with Myriad, Stratton's lab, and many others of the world's best plunging into the fray. "I have to say that we were not—certainly, I was not—optimistic as to the likelihood of our chances in this race," Stratton says. "Myriad had just identified BRCA1. And they were very experienced in the art of finding genes by this process known as positional cloning. They were very well resourced and well financed." Still, Stratton's lab forged ahead. "If nothing else, we would learn from the experience of doing it," he says.

Geneticists the world over sent Stratton DNA from their breast cancer families. They'd narrowed down the relevant part of the chromosome to about one million base pairs, a region that could encompass from one to thirty different genes depending on how gene-rich that chromosome happened to be, something that was unknown at the time, as the Human Genome Project was not yet complete.

Duke University, the Institute of Cancer Research in London, and Stratton's Sanger Centre collaborated, each doing a component of the research. Stratton is particularly proud of his group's work—they used a new method of DNA sequencing to burnish those one million base pairs to a highly accurate shine, and released the data publicly. "That was the first million bases of DNA in the human genome to be finished," Stratton says. The method used is now known as Sanger sequencing.

"As time went on, we became a little more optimistic," Stratton recalls. "There were some bad moments—we were phoned up twice by major journals who said they'd heard a rumor that the BRCA2 gene had been found. That was demoralizing." But demoralizing or not, the rumors were just that—rumors.

The lab worked at full tilt, collaborating and laboring long hours. "We didn't waste a drop of energy," Stratton says. "Everybody really committed themselves to delivering this huge body of work over a short period of time. Great drive, great commitment, great determination. . . . It was a wonderful time to be there because of that sense of commitment to a single common purpose that they could really treasure. To be part of that atmosphere was a real privilege."

That atmosphere lasted right through the day Stratton came to work and found his post-doc, Richard Wooster, waiting with some sequencing results. Together they worked their way through the genetic code until they found exactly the sort of mutation you would expect in a BRCA2 breast cancer family. "At that moment, it looked like we had landed the gene," Stratton says. He felt excitement, of course, but also pure awe. "It was that feeling of you've gotten up to the summit and you're looking down the other side and that's amazing. Looking down the other side didn't reveal anything we could interpret," but still. He and his team also felt "a sense of the impact this would have in the world. We felt very humble and we also felt slightly petrified for all sorts of reasons."

The anxiety of "Will we find it first?" was replaced by the anxiety of "Will someone else publish the results first?" In the coming weeks, Stratton's team discovered several different mutations in this gene in different families that confirmed the presence of BRCA2. The group took out patents in the UK and Europe to stake the claim to free use. "General consensus was that we would patent it for defensive reasons," Stratton said. "We would patent it and license it freely." In 2013 he would be knighted for this discovery and for his many other contributions to cancer research.

Stratton's group published their results in *Nature* on December 21, 1995, and the nonprofit Cancer Research Campaign, which had funded their research, filed for a patent in the UK. The day before Stratton's group's paper went to press, Myriad announced to the news media that it also had the BRCA2 gene, had deposited the sequence into a public government databank that day, and had filed for a US patent. The company would publish its slightly more accurate and complete results in early 1996. The genetic race was about to move from the scientific realm to the legal one.

Though we've come a long way from my mother's plea to her friend to fund research seeking the BRCA mutations, there is still plenty to learn—even in 2014, there are still some cancer families whose disease is not attributable to BRCA1 or BRCA2 mutations, or to any of the other cancer-causing mutations that have been discovered, so there may be some as-yet unknown genetic mutation or interaction at work. And of course, BRCA1 and BRCA2 are both very large genes. Just as there are more ways to misspell "antidisestablishmentarianism" than there are to misspell "pants," so too are there lots and lots of different ways the large BRCA1 or BRCA2 genes can go wrong. Some misspellings are harmless, and some are cancer causing, and of course, we aren't sure where many mutations fall on that spectrum. A battle has been won, but not the war.

4 | Myriad's Monopoly

There may be fifty ways to leave your lover and more than one way to skin a cat, but in the United States, before 2013, if you wanted to know whether you carried your family's grim legacy of breast and ovarian cancer, there was only one way to find out: pay Myriad Genetics—or get your insurance company to pay—upward of $3,000 for DNA sequencing, a process that my husband's scientist friends assured me they could do in a lab for a few hundred dollars, tops.

At first, it seemed like an indignity to me, for Myriad to own a patent on something that had been in my family—quite literally inside our bodies—for multiple generations. I felt outraged, as if some bureaucrat had casually strolled inside my house and announced that I no longer owned that nice antique teacup my grandma gave me.

It took my father, a retired tax attorney who once litigated on behalf of pharmaceutical companies, to help me see the other side. Myriad sank huge amounts of money into the problem of finding the BRCA1 and BRCA2 genes, not knowing, necessarily, what it was going to find or how long it would take, creating its own roadmap of markers and refining algorithms and methods of sequencing DNA in order to pinpoint the genes. What motivated all this investment in learning about a condition I carried? he asked. It was the yearning for profit, of course. Didn't Myriad deserve to recoup its investment? Didn't the company deserve a payoff for all its hard work?

As I followed the Supreme Court case against Myriad, my father's words rang in my ears. Issues that were easily and clearly answered did not make their way to the nation's highest court, he had said. There was legitimate disagreement here.

In order to understand it, it's necessary to know a bit about how patents work. Our current patent law dates from 1793 and describes patentable material as "any new and useful art, machine, manufacture, or composition of matter." An act passed forty years later created the Patent Office, and now patent law has restricted patentable subject matter to "nonobvious" and "useful" discoveries—no patenting spoons or washing machines that don't successfully wash clothes. The courts have interpreted patent law to limit patentable subject matter, according to the US Patent Office website, and it is not possible to patent "the laws of nature, physical phenomena, and abstract ideas." So stuff like gravity, rocks, and mathematical formulas for the meaning of life are strictly off the table. Ever since a law passed in 1995, patents may now last for twenty years from the filing date, with patentees paying renewal fees at four, eight, and twelve years to keep the patent for the full time.

Patents reward disclosure with intellectual property rights. The very word "patent" comes from an older Latin verb *patere* meaning "to be open." In exchange for describing their inventions and making them available to the public—as opposed to squirreling them away as trade secrets indefinitely—inventors earn the right in the United States "to exclude others from making, using, offering for sale, or selling the invention throughout the United States or importing the invention into the United States." That's important—a patent doesn't give you the right to make or sell something. Rather, it gives you the right to kick competitors out of the market for up to twenty years. "Think about it like a fence," Kevin Noonan tells me. "I don't have to own the property, but if I have a fence that keeps you from coming on . . ." Noonan is a lawyer with a PhD in molecular biology and is the founding author of the *Patent Docs* blog, which focuses on biotechnology and pharmaceutical patent law. He also filed a friend-of-the-court brief supporting Myriad's position. Patents, in other words, control how people can access an invention, honoring a creator's hard work with the potential to nab royalties. After the patent expires, the knowledge passes into the public domain.

In a way, the question of whether human genes should be patentable is an argument about obsolescence. As Noonan points out to me, in 1994, when Myriad found the BRCA1 gene, that represented a huge technological undertaking. It took scientists nearly two decades to identify and locate it. The company did BRCA2 in less than a quarter of that time. He asks me to think about how much cell phone technology has advanced since the mid-1990s, conjuring images of clunky car phones versus the sleek pocket computers we carry today. So, too, has DNA sequencing technology advanced. The testing methods Myriad described in its patents "are really a little horse and buggy compared to what people are doing today, and that's really not surprising," Noonan says. No wonder my husband's buddies could sequence my DNA for a fraction of Myriad's cost—it's way easier to get text messages on an iPhone than on a car phone. Noonan also points out that the Human Genome Project wasn't completed until 2003 and that the twenty-year window on patenting means that, as Noonan puts it, "By 2018, 2020, all the gene patent claims, whether they've been granted or not, are just going to be expired by the way the law works. . . . We've kind of moved past the gene age in that sense." A charitable view of gene patenting might simply suggest that what we pay—the thousands of dollars that BRCA women and men shell out to have their blood tested—is simply a temporary state of affairs, the royalty that we donate to future generations who will benefit from the knowledge after the patents have expired. A less charitable view would argue that genes aren't patentable because they're part of nature and that restricting access to them hinders meaningful scientific research and penalizes women who take genetic tests by preventing them from getting a second opinion.

In addition to the legal issues at hand—are genes patentable?—the Myriad court case filed by the American Civil Liberties Union (ACLU) and the Public Patent Foundation (PUBPAT) against the company on behalf of a consortium of researchers, physicians, and individual women includes an emotional, commonsense gut appeal. Though ACLU attorney Chris Hansen is now retired, his job used to be to seek out and litigate new cases. When the organization's full-time science adviser told him that human genes had been patented, he told her she must have gotten it wrong. "She said, 'No, the genes themselves are patented,' and I said, 'That can't be right,'" he tells me. Although he was an experienced litigator, he knew little about patents or genetics at the time. On a human level, though "it just seemed really obvious to me," he says, that human genes couldn't be patented. He spent the next two years interviewing experts "to see whether the commonsense reaction had any traction in law."

After concluding that it did, the ACLU began to put together a lawsuit. Although many human genes had been patented, the ACLU settled on litigating against Myriad based on the BRCA genes for several reasons. For starters, Hansen explains, "Breast cancer is about the most recognizable disease that exists. We knew the case was going to be controversial; we knew it was going to draw a lot of attention." Perhaps public sentiment might give them the edge. "Everybody in the country has been affected in some way or some form by breast cancer, and we thought it would resonate with judges and the public," he says. On top of that, according to ACLU cocounsel Sandra Park, Myriad Genetics had aggressively enforced its patent rights, maintaining a monopoly on providing tests to patients. She argued that Myriad's patent enforcement had "a chilling effect on researchers in the field"—a point Myriad would contest. In the end, she says, it was two factors, "the medical significance of these genes and Myriad's aggressive enforcement of these patents," that drove the ACLU's suit.

Truth be told, Myriad does have a bit of an image problem—odd considering that the company has done things like helping to identify remains from the World Trade Center bombings and discovering a number of hereditary cancer genes that help save lives, actions that most people would consider noble. I asked patent expert Dr. Robert Cook-Deegan, a medical researcher, biologist, and professor at Duke who worked in Washington for twenty years, including stints on the Biomedical Ethics Advisory Committee, the National Center for Human Genome Research, and the National Academy of Sciences, why. In addition to recounting a little of Myriad's history of suing people who infringe on its patents—as the company is well within its rights to do—he tells me that one of the reasons "that there is so much pent-up hostility to the company" is its way of dealing with its clientele. "It doesn't help that they have a product that's for women in their thirties to their fifties . . . and all of their spokesmen from time immemorial are white men in their fifties and sixties," Cook-Deegan says. Mark Skolnick is undoubtedly a brilliant scientist, but for all his accomplishments, he's not very likeable. When filmmaker and BRCA carrier Joanna Rudnick interviewed him for her film *In the Family*, he came off as, well, an arrogant mansplainer, denying that there was any controversy around the gene patents. "There's no controversial patent. It's all very easy to understand if you take the time," he tells her, as if to imply that this woman who wants to interview the guy who found BRCA1 has probably not taken the time to try to understand the issues. And of finding the BRCA genes, he tells her, "We did it to win the race. The group that was in that laboratory working seven days a week were doing it to beat the other team, period. And we won." I guess I'm glad that the genes were found—whether to beat another team or, you know, because women are real humans who deserve to not die of cancer from generation to generation. I mean, the important thing is that his team found the genes. Still, these sorts of public slipups are not going to win Skolnick or Myriad any Ms. Congeniality prizes.

It's easy to demonize Myriad as the haunted house on the hill, the big scary corporation coming in and enforcing its mean will on women's bodies, but that's not a fair cop either. There's no rule that says innovators have to be selfless and cuddly—isn't it enough for them to do the science that uncovers the underpinnings of diseases? Who among us, having discovered a gold mine, or an untapped oil field, would truly give up that tremendous source of wealth and power to the public interest?

And so, the ACLU and PUBPAT's lawsuit against Myriad had two components—one was a legal question of whether human genes are patentable subject material, a battle fought in the courtroom. The other, intertwined issue concerned whether it was ethical and right to patent human genes, a question largely tried in the court of public opinion, but one that influenced the legal issues at hand.

Let's take the ethical question of gene patenting first.

On the "it's not ethical" side, we have the ACLU and PUBPAT, who argued that patenting actual DNA sequences impedes data sharing and research important to the future of bioinformatics. They noted that Myriad sent cease-and-desist letters to some researchers in connection with the patents. If DNA sequences aren't free, then how can laboratories conduct research on the genes in order to unravel their functions and find therapies? Attorneys for Myriad would cite thousands of papers on BRCA1 and BRCA2 in their arguments in rebuttal. Still, according to cocounsel Sandra Park, Myriad doesn't have to prevent research in order to harm the development of science. The point, she tells me, is that Myriad could choose to sue researchers if it feels like it; if a researcher's ultimate goal is clinical applications, "and you know that Myriad will likely go after you at that stage, there is a chilling effect that goes on at the front end." Would researchers choose to study these patented genes, knowing that they might have their discoveries taken from them? That's one ethical quandary.

Another quagmire lies in the fates of researchers who perform clinical studies on BRCA patients. According to Dr. Robert Cook-Deegan, the Myriad patents extended patent protection to a diagnostic test, which means that, for a patent holder, "you in effect become the arbiter of the standard of care for doing that test because no one else can do it." According to two doctors who provided affidavits in the lawsuit, Myriad didn't permit researchers to tell their subjects the results of their BRCA1 and BRCA2 tests, forcing them to violate accepted medical ethics.

Finally, Myriad's monopoly on BRCA testing, founded on its gene patents, created problems for patient access, as professor Lori Andrews of the Chicago-Kent College of Law, who specializes in the legal issues around medicine, points out to me. She also chaired the federal Working Group on the Ethical, Legal, and Social Implications of the Human Genome Project and authored an amicus, or friend-of-the-court, brief supporting the ACLU and PUBPAT. "I received grants from foundations and federal agencies to look at the impact of gene patents and what I found was horrifying," she says. "Interfering with health care—even someone like Angelina Jolie, even wealthy people could not get a second opinion before cutting off their breasts." And of course poor women or women without Myriad-approved insurance might not get access to genetic testing to begin with. The plaintiffs also argued that the tests didn't reveal all known mutations and didn't use techniques that could do so, meaning that the tests gave false negative results to people who really did have harmful BRCA mutations. One study published in the *Journal of the American Medical Association* stated that about 12 percent of patients from high-risk families received such false negative results. Of course, it's unreasonable to expect Myriad to test for a mutation that it is not aware of—it can't test for things it doesn't know exists. Still, if the DNA itself hadn't been patented, the plaintiffs contended, other labs could offer more comprehensive testing using newer techniques, including testing tumor specimens preserved in paraffin from deceased family members, which Myriad did not regularly perform. Myriad's foes also argued that a lack of independent testing prevented the scientific community from performing research on variations of uncertain significance—alterations detected on BRCA tests that might or might not affect cancer risk. Myriad argued that it permitted some labs, for example at Yale and the University of Chicago, to perform confirmatory testing under narrow licenses but that second opinions wasted resources unless there was some doubt about the original test. As for the original test, the company argued that its testing is the gold standard and that it is continually updating and improving its tests, including efforts to determine the clinical significance of VUSs (variations of uncertain significance), which has resulted in the clarification of 850 VUSs from twenty-one thousand patients.

Andrews also points out that, as genetic sequencing becomes easier and cheaper, the patents become more cumbersome. "We're moving toward the thousand-dollar genome test," she says. We couldn't do that if every gene had a royalty on it. If patients had to pay even a very modest fee, say $10 for each gene in the genome, she explains, that would add $220,000 to a patient's test. Similarly, the plaintiffs argued that Myriad's monopoly meant that it could control access to the test financially by setting the test's price and deciding what insurance it would accept. Myriad, as it turns out, has a program to offer free testing to uninsured patients, as well as offering a no-interest financing program to allow patients to pay out-of-pocket costs over up to twenty-four months.

On the other hand, the scale of Myriad's low-income help, in the form of free and low-cost tests, compared to its profits is difficult to uncover. In 2012 Myriad told the Patent Office that some one million patients had been tested for BRCA, and that over the last five years it had provided free tests for some four thousand patients, a comparatively small number. Those four thousand tests had a value of about $13.3 million, if each of the patients would have otherwise paid the current list price of $3,340. Compared to the more than $2 billion the company has raked in over the last fifteen years, $13.3 million over five years doesn't sound like a whole lot. A 2014 opinion by a judge in another legal matter gives the much higher number of thirty-five thousand patients helped over the last five years but lumps together free and reduced-cost testing, as well as Myriad's interest-free financing programs.

That's the anti–gene patenting ethical case in an extremely reductive nutshell: limiting scientific research and failing patients by creating cost barriers and preventing the rivalry that might make companies vie to create more accurate and cheaper tests.

To make an ethical case for gene patenting, I turned to Noonan, since despite my numerous requests over several months, representatives from Myriad said they were too busy to talk to me. Though Noonan wasn't directly involved in the case, he follows patent law like a hawk and filed an amicus brief supporting Myriad's position. He reminded me that the methods used to sequence BRCA genes were—at the time of the genes' discovery—much more expensive and cumbersome than they are now. He asks me to imagine a world without Myriad. Let's say the genes are discovered anyway, and the person who does so publishes their sequences but doesn't patent them. In all likelihood, the test will still be available; the question is whether doctors and patients will be informed about its existence and whether insurance will cover the diagnostic. Noonan argues that Myriad, because it had a business interest in spreading the test far and wide, was able to make it more widely available. As a business, it devoted time and resources to explain the significance of its findings to loads of ob-gyns, it argued with insurance companies to get them to cover the test, and it talked to state legislatures. "It's not because they're altruistic, it's not because they're wonderful people, it's because they want to make money," he says. Some people "might say that sucks," he admits, but well, "that's capitalism." In our capitalist system, profit provides the motives. Implied is the idea that if we remove gene patenting, if we don't provide companies with intellectual protection to invest large sums of money in spreading lifesaving research, they might not do it. Without patent law and the profit incentive, the BRCA test might not have received the marketing and insurance coverage needed to reach many patients.

As a scientist, Noonan also recognizes some larger implications for gene patenting. "BRCA is really an anomaly because most cancers don't develop from one mutation" but rather develop from the complex interaction of various genes and environmental factors, the mechanism of which is not yet fully understood. His worry is "that if you say you can't protect it [a test for cancer predisposition] by patenting," then people won't disclose their methods, and science will stall or won't move as fast as it has in the last few decades. The next cancer test might be concealed as a trade secret, just like the recipe for Coca-Cola has remained secret all these years.

There's the ethical case for gene patenting: it provides profit motives to companies to sink research money into discovering genes that affect heritable diseases and into dealing with all the bureaucratic red tape needed to make the relevant tests available across the country. It also incentivizes disclosure of discoveries. Without gene patents, maybe we wouldn't have a BRCA test at all.

In some way, the ethical case hinges on how much you believe in profit motives. If Skolnick hadn't mobilized his millions in investment capital, would one of the other labs in the race, such as King's, eventually have found and set free the BRCA mutations? And if they had, would the test have become commercially successful enough for someone to turn out a quality product without the assurance of patent protection?

The legal issues the case raised are a good deal more arcane. The ACLU and PUBPAT challenged only a handful of Myriad's numerous patents relating to the BRCA genes. According to Cook-Deegan, there were fifteen claims at issue contained in seven patents, while Myriad has over 250 claims in twenty-four patents relating to the BRCA genes. The case challenged five of Myriad's broad claims on methods for examining the genes, as well as nine claims relating to composition of matter—the DNA itself. The plaintiffs won their suit on method claims—claims on ways of comparing DNA sequences to one another—in the appeals court, and Myriad did not appeal that part of the ruling; so the Supreme Court case dealt primarily with the composition claims, the claims that related to the actual genes in isolation themselves. The core question of the case became: Are the BRCA genes a product of nature or human ingenuity?

In the time it takes for a patent case to settle, you could eat five whole hams with chopsticks—by the time a case concludes, it feels like the patents in question might have expired anyway. But still, the issue is an important one. As cocounsel Park puts it, "A lot of folks have tried to portray this as an issue that will go away—getting patents closely related to what we think is not patentable," but "Myriad is continuing to get patents related to genes that are problematic." The clock ticking on the patents in question, though, made the plaintiffs' legal strategy ballsy. Park explains, "We dreamed big from the beginning because we knew that the only way we would win is by getting to the Supreme Court." In other words, successfully trying this case was a long shot because the Supreme Court takes so few cases, and the court with jurisdiction immediately below it—the court of appeals—tends to favor patent holders. Still, the ACLU and PUBPAT tried to shoot the moon, and they succeeded. They won their first court case, lost their appeals case, and applied to the Supreme Court, which kicked the case back down to the appeals court again and asked it to decide the case a second time based on a recent Supreme Court ruling. After the second go-round with the appeals court, which yielded the same result—a ruling in favor of Myriad—the plaintiffs again applied to the Supreme Court, and this time, it agreed to hear the case. By that time, the consortium of plaintiffs had been knocked down to a single individual, medical geneticist Dr. Harry Ostrer. More than fifty individuals and groups filed amicus briefs, illustrating the case's importance to scientists, researchers, law professors, businesses, and patients.

It was a battle of analogies drawn from earlier Supreme Court cases. Is a gene snipped out of its native DNA by restriction enzymes more like an oil-eating bacterium, or a fruit whose skin has been treated with borax? A mix of bacteria used to fertilize crops, or adrenaline extracted from the suprarenal glands of animals? Cellulose refined from wood pulp, or purified tungsten or uranium? A medicinal leaf plucked off an Amazonian plant, or a baseball bat carved out of a tree limb?

Myriad, naturally, argued that the isolated DNA sequences concerned in the patents were products of human ingenuity. I mean, we can't isolate DNA sequences in our kitchens—it requires a substantial level of technical skill and equipment. Myriad's attorneys argued that their patents were for man-made molecules, because isolating a gene from its native DNA sequence created a new chemical entity, one produced by human intervention, a molecule useful in its capacity as a screening tool for hereditary cancer syndrome. They found cases that revolved around purification of natural substances to be helpful to their cause. Myriad's brief argued that "the established rule for over 100 years has been that isolates or extracts from natural materials that reflect human invention *are* eligible for patents." For example, in the 1911 case *Parke-Davis & Co. v. H. K. Mulford Co.*, the court ruled that purified adrenaline made from the adrenal glands of animals was patent eligible. If you extract the molecule that gives strawberries their distinctive taste and smell, or purify the prostaglandin hormone, or make concentrated vitamin B12 from bacteria, all these are patent eligible according to the courts, Myriad argued in its brief. The BRCA genes are simply a natural product that has been purified into a more useful state. Similarly, Myriad's brief cited the *Chakrabarty* case, in which the courts ruled that bacteria genetically engineered to consume crude oil could be patented, according to the brief, because "although the molecules were derived from natural materials, those materials in a state of nature provided none of the benefits of the claimed molecules." Similarly, the brief argues, "Without Myriad's work, there is no indication whether these particular molecules, and the valuable uses to which they have been put, would ever have come about."

The plaintiffs argued that human DNA—even isolated DNA—is a product of nature and that Myriad's isolation of the BRCA genes did not rise to the level of innovation demanded to change something from a natural product into a product of human ingenuity. They cited the *Funk Brothers* case, in which the Supreme Court ruled that mixing together several sorts of natural bacteria as a fertilizer wasn't enough to warrant a patent because "their qualities were the work of nature, not the patentee." They also cited the *American Fruit Growers* case, in which the court overturned a patent on fruit that had been treated with borax to resist mold. Even though borax-treated fruit doesn't exist in nature, the fruit's primary use—as food—remained the same. So too, they argued, isolating a segment of DNA does not change its structure or function—it still codes for the same proteins, and it's still got that double-helix shape. They also used *Chakrabarty* to support their position, arguing that isolating genomic DNA did not rise to the levels of innovation required for a patent. By the same token, they argued, the purification argument didn't work. They cited the *Gen. Elec. Co. v. De Forest Radio Co.* case of 1928, in which the court ruled that superpurified tungsten couldn't receive a patent, because in the words of the court, "Naturally we inquire who created pure tungsten. Coolidge? No. It existed in nature and doubtless has existed there for centuries. The fact that no one before Coolidge found it there does not negative its origin or existence." In the same way, the plaintiffs' lawyers argued, "Myriad did not invent the isolated DNA. Myriad did not invent any of the characteristics of DNA that are incidental to its isolation. Myriad did not invent the length, composition, or function of the BRCA1 or BRCA2 genes; human biology determined these qualities of the two genes."

The oral arguments before the Supreme Court show the justices carefully grappling with where the line between a natural product and human invention lies. Myriad's attorneys try to sell them on this baseball bat analogy—that clipping a BRCA gene out of DNA is like refining a baseball bat from a tree branch; the human mind has to create the shape of the bat before it can be carved down. The plaintiffs' attorneys try to sell the justices on an Amazonian plant leaf analogy. Plucking a medicinal leaf off a plant doesn't entitle one to a patent on the plant. Just as plucking the BRCA genes—genes decided by biology, not human invention—out of the genome doesn't entitle one to a patent on the genes themselves. And the Justice Department made an unusual appearance to advocate a middle path, namely that human DNA is not patentable but cDNA is.

cDNA is a sort of lab-created DNA that occurs in nature only very rarely. Here's a quick and dirty take on how it's made. In order to regulate cell function, DNA unzips itself and makes molecules of RNA, which is chemically slightly different from DNA and codes for proteins. The RNA then gets edited—our DNA has lots of good stuff that codes for important things (exons) and lots of irrelevant junk (introns). During RNA splicing, the introns get edited out until only the exons—the important protein-coding stuff—remain. In a laboratory, scientists can then take this finished RNA molecule and create a strand of cDNA from it. The resulting cDNA differs from the original DNA in that it has the irrelevant stuff edited out.

The Justice Department argued that cDNA may be patentable because it's genetically engineered, just like the oil-eating bacteria in *Chakrabarty.* Cocounsel Park tells me that it's unclear whether Myriad's patents covered cDNA and points out that the government's position was that cDNA is not a product of nature. Even if that is true, cDNA may still not be patentable on the grounds that it is obvious, but in this particular case that was not at issue. Noonan says the Justice Department argument gave the Supreme Court "a nice simple sheep-and-goats dividing line that would theoretically make everybody happy."

The Supreme Court had to balance the considerations of facilitating free scientific research with the concerns of industry—not just from an economic perspective, but because pharmaceutical companies do good stuff like isolate natural extracts and turn them into, among other things, aspirin and chemotherapy drugs. The justices unanimously overturned Myriad's patents on genomic DNA while leaving the door open to patents on cDNA, and their brief largely followed the Justice Department's logic. Isolated DNA is a product of nature and thus unpatentable, while cDNA is the product of significant human creativity and may thus be patentable.

As far as the impact of the decision, both Noonan (who authored an amicus brief supporting Myriad) and Cook-Deegan (who authored an amicus supporting neither party) agree that the immediate impact will likely be modest. Noonan calls the ruling a "narrow" and "split-the-baby" decision that gave something to both plaintiffs and respondents. Cook-Deegan points out that to his knowledge only two companies, Myriad and Athena Diagnostics, currently operate on a service-monopoly model relying on gene patents. "Pretty much everybody else isn't going to do that," he says. He thinks the decision will change how people write their patent claims, but that it won't fundamentally change "the kinds of patents that really matter in the real world of biotech." Grotesquely broad patent strategies will be reined in, but that's it.

Still, in a legal system that operates based on precedent, Noonan argues, the implications of a Supreme Court decision don't merely lie in the matter of law the case decides but in how that case is later used by other litigants. He expects the decision to ripple outward. "Litigants being litigants, and lawyers being lawyers," he tells me, he believes that the Myriad decision, while narrow, "will be called to cover lots of things, and then the question is how successful will those efforts be?" Already, he says, the case is being cited in a stem cell suit that does not pertain to genetics.

In the meantime, the bottom line may not change all that much for BRCA patients. Myriad still has its vast database of genetic mutations that it can use to predict particular risk. Not all BRCA1 and BRCA2 mutations necessarily connote the same level of risk, because some genes are more penetrant than others. "Penetrant" is another way of talking about probability. When a gene is highly penetrant its effect will show up in a large number of people with that gene. So, some BRCA1 and BRCA2 mutations are highly penetrant, meaning that most of the people carrying those mutations will get breast or ovarian cancer. Other BRCA1 and BRCA2 mutations are not so penetrant, meaning that most of the people who carry those specific mutations will probably not get cancer, or at least, not because of the mutations. As the sole provider of the BRCA test, Myriad has more data points when it comes to estimating customized disease risk. Cook-Deegan points out that this is a double-edged sword. Normally, he says, such scientific results would be validated independently, but since Myriad's database is proprietary, "We don't know how valid the interpretations are. There is no reason to doubt them, but there is equally no reason to believe them because they aren't published and they aren't publicly available." ACLU cocounsel Park concedes that Myriad does "have an advantage because they have the largest database about the genes and they've refused to share that information," but she argues that "from the experts I've spoken to, that advantage should be considered in the right context." She believes that the database is mainly an advantage for very rare mutations. "But for most patients, that's not going to matter for them," she explains, and over time the advantage should dissipate as more labs develop their own databases. Noonan, on the other hand, points out that the patent suit has made Myriad less likely to share its results, possibly penalizing scientific progress by disincentivizing sharing. "They are very personally offended by all this," he says, comparing the situation to the story of the little red hen who worked so hard to make the bread. "They did all this stuff, and at the end of the day they get the bread, and now everyone wants a piece."

Finally, the Supreme Court decision only knocked down a few of Myriad's many patents on the genes, leaving many additional claims around methods and uses. Shortly after the Supreme Court's decision in the summer of 2013, Myriad brought suit against two companies, Ambry Genetics and Gene By Gene, that declared they would begin offering BRCA testing. Myriad sought an injunction to prevent the companies from opening up shop on the grounds that this would infringe on several of Myriad's patents on methods of looking at the BRCA genes, particularly on its methods of identifying BRCA mutations. Its opponents say these methods are known and obvious techniques. By early 2014 half a dozen companies had been attached to the suit, though Gene By Gene and Myriad settled out of court. In March 2014, a judge shot down the preliminary injunction, but the case could still go to trial or get settled out of court, depending on how the participants proceed. Park tells me that in her opinion Myriad's suit is "a strategy and business decision," aimed at squeezing out "a last few months of monopoly. They make a huge profit on testing." Some of the companies also filed their own countersuits against Myriad.

Now, at the time of writing, there is more than one way to find out if one has a BRCA mutation—several companies have opened up shop, and patients have multiple options. According to Joy Larsen Haidle, president-elect of the National Society of Genetic Counselors, the decision has changed the practice of genetic counseling. There are many genes that can influence cancer risk. "We needed to test for BRCA1/2 first, and once that was negative we could look at other, less common breast cancer genes," she says. Now they can do multiple gene panels from the start. The decision "has changed things a lot, and we're starting to identify inherited risk factors in families we would not have identified previously," she says. Still, the legal wrangling hasn't yet reached its conclusion, and depending on how the case shakes out, BRCA patients could still end up back at square one.

5 | Positive

I learned my mother's genetic results on an ordinary spring day while leaving my local subway station on the way home from work. As I walked through the supermarket parking lot, cell phone to my ear, we whipped through the pleasantries—instead of her usual questions about my day and recent cooking activities, she cut to the chase. "I'm positive," she said.

There wasn't much to say after that, so we hung up. On the other side of the parking lot, I stood on the corner, stunned for a moment. I didn't know how to react. It wasn't a surprise, exactly, that a woman who'd had estrogen-negative breast cancer at thirty shared her mother's BRCA1 genetic mutation. Yet a negative test on her part would have spared me the drama of continuing further down this path. My risk of carrying a BRCA1 mutation was now 50 percent. I had a 50 percent chance of having what was then believed to be a 40 to 87 percent chance of developing breast cancer during the course of my life. Did that mean my known risk was now 20 to 43 percent? Could you multiply probabilities like that? Why hadn't I taken statistics in college?

The number representing my risk worried me because I wasn't living my life perfectly. I drank beer in the evenings with friends after work or class. Occasionally, I'd bum a cigarette, relishing the feeling of breaking the rules. Because of Boston's awesome mass transit system, I walked a lot, but I didn't go to the gym often. What would I do if I had the gene? Would I give up my hedonistic pleasures? Would it even matter enough to counteract my risk?

At home I delivered the news to my boyfriend, George, and to Chip, our close friend and one of our four roommates. I don't recall crying, but I do remember wanting something, anything, to dampen the swell of panic that rose in my throat. I felt one step closer to the grim family legacy.

Over the next few months I scheduled a flurry of doctor's appointments, indistinct, but pervasive. Armed with my mother's test results, I revisited the genetic counselor's office in the ominous oncology department. The same calm woman sat across from me, speaking in the same calm, hushed, your-puppy-just-died tones. I didn't have to take the test at all, she told me. And if I did want to take it, that didn't have to be now. Out of an abundance of caution, though, they'd put me on the screening protocol. "We're going to treat you like you have the gene until you prove to us that you don't," she said. I felt reassured. They would err on the side of caution. I liked caution. That meant MRIs alternated with mammograms every six months, plus extra pelvic exams—a ritual test for every season to ward off the evil eye of cancer.

The yearly medical gauntlet sounds easier than it was. For starters, there were so many appointments to schedule; I spent a lot of time shuttling here and there to have different sorts of beams shot into my breasts. Emotionally, it was difficult too. A trip to get a mammogram is nothing like a trip to the dentist. When the women in your family get cancer, you drag baggage into those waiting rooms with you. The ordeal of actually being scanned is second only to the anxiety of waiting for the results to come back to you at a later date, no matter how well you know that statistically, at your age, even with your family history, you're pretty unlikely to come down with something lethal.

But maybe, just maybe, all these scans, beams, pokes, and proddings were unnecessary. Maybe I had lucked out and avoided my family's BRCA mutation. After all if one parent—mother or father—has a BRCA mutation, there's only a 50 percent chance that their kid does too. That meant I had a 50 percent chance of being free from the family curse. Part of me wanted to take the test and get the inevitable over with, and part of me wanted to wait for the right moment. If I tested negative, I wouldn't have to worry so much, but if I tested positive . . . well, I wasn't sure exactly what I would do then, but it seemed like it'd be good to know. At the same time, I had a high level of paranoia—once this knowledge is out there, it's out there and open to misuse. What would positive results mean for my health and life insurance or for my employability? I'd heard that some women had paid for their tests out of pocket and used pseudonyms to avoid letting their insurance companies know. At my very first trip to the genetic counselor's office, I had hesitated before signing in, worried that the simple act of writing my name could somehow lock me into an eternity marked as a genetic mutant.

My worries were not unique. Although few cases of genetic discrimination had been filed at the turn of the twenty-first century, the public fear of the misuse of genetic information exposed a gap between the pace of scientific advancement and the law, brought on by the international Human Genome Project. The project, begun in 1990 and completed in 2003, aimed to sequence and identify all of the genes in human DNA.

Scientists and legislators approached genetic advancement with unbridled optimism. To put it in 1990s terms, the future was, like, totally now, dude. In a 1997 statement advocating for a federal law to protect genetic privacy, the Department of Health and Human Services wrote of the promise of unlocking the genome in understanding, treating, and preventing disease: "At one time, such medical clairvoyance seemed like science fiction. But not anymore." President Clinton shared that optimism. During a speech on genetic privacy a few years later, he said, "We can now only barely imagine" the wonders of treatment that sequencing the human genome would produce, adding, "It will transform medical care more profoundly than anything since the discovery of antibiotics and the polio vaccine, I believe, far more profoundly than that." The head of the National Human Genome Research Institute went even further, declaring in a 2000 speech that before the end of the decade, it was quite possible that "each of us may be able to learn our individual susceptibilities to common disorders, allowing the design of a program of effective individualized preventive medicine focused on lifestyle changes, diet, and medical surveillance to keep us healthy," and hoped that it would usher in an era of healthcare focused on "maintaining wellness instead of relying on expensive and often imperfect treatments for advanced diseases." Public figures made genetic research sound like a panacea, belying the complications that would arise around surveillance and the psychological inevitability of knowing your disease risk. What if "maintaining wellness" meant carving out body parts?

And yet, the nervousness around genetic privacy and its potential misuse tempered the optimism of the moment—if the province of science fiction had become reality, perhaps we needed to prevent a science-fiction dystopia. One thing's for sure: the public was really worried. As early as 1993, an NIH report on the implications of human genome research fretted that if people thought genetic information might be used to deny them health coverage, then they would forgo tests that might otherwise help prevent disease, encourage early treatment plans, or help them plan for the future. A report five years later from the US Department of Labor confirmed this suspicion—that fear of discrimination might drive people away from using the new technology. The report cited a recent poll showing that 85 percent of the general public was very or somewhat concerned about insurance companies and employers having access to genetic information; a telephone survey discussed in the report found that 63 percent of people wouldn't take genetic tests for disease if their bosses or health insurance companies could learn the results. The Department of Labor report also referenced a multiyear study of Pennsylvania women at high risk for breast cancer—almost a third of the women invited to participate refused because they feared discrimination.

As with so many things related to BRCA, the fear seemed worse than the actuality—the public's worries about genetic discrimination vastly exceeded the incidence, but they weren't entirely unfounded. In 1998, some employees of Lawrence Berkeley Laboratory won a lawsuit against their employer for testing their blood and urine samples for sickle cell trait (a genetic test), syphilis, and pregnancy (nongenetic tests) without their knowledge or consent. The California appellate court called the performance of such medical tests without consent "the most basic violation possible" and sided with plaintiffs on grounds of the Fourth Amendment, which protects citizens against unreasonable search and seizure. The US Equal Employment Opportunity Commission (EEOC) brought a similar suit against the Burlington Northern Santa Fe Railway Company in 2001, alleging that, without asking, the company tested employees who filed for worker's comp due to carpal tunnel syndrome for a deletion on chromosome 17 that makes people more susceptible to wrist injuries. The lawsuit settled, with the railway agreeing to the EEOC's terms.

Legislators acted because the public feared for its genetic privacy, and this disproportional fear diminished the utility of the new knowledge—people who could benefit from testing turned it down—and impeded the progress of medicine, which depends on willing research participants. Finally, legislators feared the collision between genetic information and racial discrimination; they remembered the sickle cell anemia debacle of the 1970s. In 1971, President Nixon dedicated $10 million to programs geared toward the disease, a blood disorder caused by a recessive genetic trait carried largely by people of African descent. A year later, he signed the National Sickle Cell Anemia Control Act, which dedicated a further $15 million to screening and counseling programs, education, and research. Money to treat and research sickle cell anemia—what's wrong with that? But in the wake of the funding increase, some states began requiring African American children to undergo mandatory genetic screening for the trait before entering school, a discriminatory practice that targeted individuals based on race. Legislators recognized that genetic discrimination had the ability to overlap and reinforce racial discrimination—it raised the ugly specter of eugenics, and certainly no one wanted that.

After the discovery of the BRCA mutations, researchers uncovered a high rate of certain BRCA1 and 2 mutations in people with Ashkenazi Jewish heritage, leading to similar fears. As journalist Jeff Wheelwright writes in his book about the genes, *The Wandering Gene and the Indian Princess*, after these discoveries, "some Jewish leaders advised people not to participate in genetic studies any longer. This is what we get for helping? they asked, linking concerns about Jewish genetic defects to Nazi racial classifications and the Holocaust. . . . Still, with very few exceptions, the genetic abuses that critics warned about haven't materialized and the dehumanizing scenarios remain very far in the future." The potential for abuse is understandably frightening, even if it has not been realized.

Federal laws against genetic discrimination were a long time coming, with members of Congress introducing bills as early as 1995. In

2000, President Clinton signed an executive order prohibiting genetic discrimination in the hiring of federal workers, and eight years later, President Bush signed the Genetic Information Nondiscrimination Act (GINA) into law. GINA extended the protections of Clinton's executive order to all citizens, preventing employers from collecting information about employees or applicants and preventing health insurance companies from using genetic information to charge higher premiums. Although by this time most US states had passed laws about genetic discrimination, the laws had varying scopes; GINA established a minimum level of protection. The bill passed the Senate 95–0 and the House 414–1, with Ron Paul the lone dissenter.

The new law became effective in 2009 and essentially made it illegal for employers or insurers to collect genetic information and family history unless they make clear that such collection is voluntary and will have no effect on an individual's eligibility for employment and for group or individual health insurance, and no effect on their current status. Still, GINA does not cover long-term care insurance, disability insurance, or life insurance.

Since the law became active in 2009, the complaints of genetic discrimination to the EEOC have increased by a modest number each year, with 201 charges filed in 2010 and 333 filed in 2013. Since the law is young, it's unclear whether this represents an actual increase in genetic discrimination or whether it's simply a product of people gradually becoming aware of the new law and using it. The EEOC mediates disputes between employees and employers but, with rare exception, does not file lawsuits. Over the course of four years, it heard a total of 1,059 charges of workplace discrimination involving GINA, not a lot considering that we're a nation of 317 million people.

In the ensuing months after my mother received her results, I thought carefully about whether I wanted to know what lurked in my DNA. The discrimination issue gave me pause, because if I could keep my results private, I'd definitely want to know. That a family mutation explained the curse that had dogged my relatives for years put a shape to our suffering. It wasn't just some sadistic god, bent on inducing maximum terror in my relatives; we simply had a piece of erroneous DNA. And I had a one-in-two chance of avoiding the pain my family suffered. Oh yes, I wanted to know. At the same time, I knew receiving my results would hit me hard. Of course, I hoped to be negative, that all these mammograms, MRIs, and pelvic exams had been unnecessary—but I also thought it wise to prepare for a positive result, which I knew would devastate me. It would be worse than the limbo where I already lived. So I asked myself: Can I afford an emotional meltdown at this moment in my life? George and I had just gotten engaged. In a few months, I'd finish my MFA in fiction writing and would move from Boston to New York to start an intense journalism master's program at Columbia, which by all accounts would involve sixty-hour workweeks and countless all-nighters for the next academic year. That felt like a lot of major life events in a short time span. I wasn't sure I could endure them if I had a positive test result. And it felt like too much to ask of someone, to confront this stuff at twenty-five.

So for the next two years, I lived in this level of limbo. Mostly, I stayed busy. I learned to walk up to strangers on the street and talk to them. For class, I wrote short reported pieces about health care, firemen, welfare, and food stamps in one of the poorest neighborhoods in the Bronx. I learned to shoot video and code HTML and worked hundred-hour weeks, often sleeping in the multimedia lab the week before deadline. I researched pickles and adult make-believe for my longer projects. After graduation, I lost the internship I'd landed at a local newspaper because the industry was collapsing and they no longer had the budget for me. I scrambled to find a new job and scored an early morning shift at the news site the *Daily*

Beast. Meanwhile, I reported for my first book, landed a literary agent, and planned my wedding. Periodically I visited websites for the club I desperately wanted to avoid—BRCA mutation carriers. What I read there scared me. Women talked about their breasts and ovaries like they were weapons of destruction, "ticking time bombs" that required surgical disarmament, or assassins that deserved a murderous preemptive strike. I couldn't relate to that feeling at all. I loved my breasts, or at least, I liked them. They were part of me. I found the sentiment "My breasts are trying to kill me" off-putting, threatening, an indication that these women didn't understand my state of mind. I wanted nothing to do with them; I had no desire to join their macabre club. So of course I found a reason to criticize their words, their fear, their responses, their way of emotionally distancing themselves from the body parts that needed to go. I was nowhere near ready for a mastectomy.

After GINA passed in 2008—the year I finished my graduate education—it felt like one barrier, one excuse for not testing, had vanished. The bill became law in May, though the provisions wouldn't take effect until 2009. I stopped imagining potential fake names for myself and calculating how much it might cost to pay out of pocket and keep my insurance company in the dark. With genetic discrimination off my worry list, I found other reasons to delay testing. George and I set our wedding date for early 2009, and since he'd have to take qualifying exams for his PhD soon after, we'd decided to delay our honeymoon until May. I wanted to grab a few last months of unmarred happiness, to enjoy the wedding and the budget trip to Italy we'd finagled without emotional complication. So I waited.

At the same time, my nervousness grew. I was dancing up to the age that women in my family began to get cancer. Though time had muddled Trudy's story, some accounts suggested she'd started developing breast cancer around twenty-seven or twenty-eight. I was now twenty-seven. In the rush of graduation and job hunting and research, I hadn't been great about getting my mammograms or massaging my breasts for lumps. I didn't want to think about it, didn't want to interrogate my body for fear of what I might find. Still, I thought of cancer often.

We had a beautiful ceremony and an even lovelier fortnight in Rome and Florence four months later. I was out of excuses.

I found an oncologist, a kind man with a ready smile and a firm but not-too-firm handshake who seemed able to deliver the facts without doing so brusquely or condescending to me. I met him once, and then on my second visit, he drew my blood. He'd asked me what I'd do if I learned I was positive, and I'd vaguely replied "lifestyle changes," thinking perhaps I'd drink less or exercise more or eat less meat. I felt I had to know, though, whether my worry was reasonable or whether I'd been an unknowing hypochondriac.

He said it would take two weeks to get the results of the simple DNA test performed on my blood. My mother had done the hard work all those years ago—technicians had sequenced large sections of her DNA in order to figure out whether she carried one of the many cancer-causing mutations. For my test, the scientists at Myriad Genetics would simply perform a spot-check of my DNA to see if my gene matched my mother's. Since my father's family doesn't have a strong cancer history, we assumed his genes were fine. My mother had one broken gene and one normal one and could have passed either on to me, so I had a 50 percent chance—a coin toss—of having two normal copies instead of one with the cancer-causing BRCA mutation.

I spent the next two weeks of that July on pins and needles. Of course I wanted the test to come back negative. I'd seen the menu of options offered to BRCA women, and given a choice, I'd prefer to eat at some other restaurant. If I had my mother's BRCA mutation, I could continue intensive surveillance to catch cancer early, carve out my breasts and ovaries, or undergo hormone therapy that would put me in temporary reversible menopause but buy me a few years to make a more permanent decision. Given my druthers, I'd pick option two: not having the gene.

At the same time, testing positive felt inevitable. I was my mother's daughter, wasn't I? Dealing with the family legacy of cancer felt like part of who I was, and if I tested negative, that legacy would be ripped away from my identity. I'd be excluded from the heritage that inscribed "Long life isn't guaranteed" onto the psyches of our family. I'd simply become normal.

I felt weird. I really, really didn't want to test positive, but a negative test would negate part of my identity, and a small messed-up part of my brain would feel like a disappointed drama queen. During the weeks I waited, a friend called, and I tried to explain my fears of having the genetic mutation. "My aunt just died of breast cancer," she told me. "If you have the gene, you'll just cut them off. So much better." In the midst of her own worse grief, she said it flippantly, in a way that suggested that removing my breasts would be a pragmatic move, as if it weren't a big deal. I hung up soon after.

My mother held out hope that I'd be negative. After so many women in our family had suffered, surely I'd be spared. "When I heard you were going to get the test," she said, "I thought, 'Oh, this is so valuable to do,' and I was really convinced that you wouldn't have it. I didn't really consider the possibility that you might have it, even though intellectually I knew it." A lifelong Lutheran, she'd prayed about it. When we spoke on the phone, I'd talk about what I would do if I were positive, and my mother would remind me that I still had a very good chance of dodging the whole issue with a negative result. She told me that if I tested negative she'd be in the car in an instant, on her way to see me, with champagne in hand. And if I tested positive, well, she'd be in the car with something chocolate and creamy.

On a Thursday in July, one of George's friends dropped by our tiny apartment to pick something up and visit for a while. My phone rang with an unknown number. I almost didn't pick it up. "You should get it," George said. "It might be those results you've been waiting for." He raised his eyebrows meaningfully. I'd almost forgotten about the genetic test that week, but now I flew to pick up the phone. It was my oncologist, so I went into the bedroom for privacy. He softened the blow right away with, "I've got the results, but I'm afraid they aren't the ones you wanted." I sat down on the bed. I had the same gene my mother and grandmother had. I felt numb, shocked. I should tell someone, I thought. I should tell George. But we had company. I tried to make my face neutral and walked into the living room. I didn't want to make our visitor uncomfortable, but I couldn't suppress this life-changing news either. I'd make myself calm. "That was the doctor," I said. "I have it. I have the gene." Saying it made it real, and my voice cracked. "You barely got that sentence out," George remembers, "because you had already started to cry." I immediately fled to the shower.

It felt safe to weep in the shower, safe to be alone, safe to be with my body for one of the last times. In my bones, I knew that eventually, ideally some day far in the future, it would be all over for me and my breasts, but I didn't want it to be. I loved my breasts. I loved having them attached to my chest rather than lying in a lab somewhere. The thought made me protective of them, and I held them in my hands. They felt soft and vulnerable and like part of me—not part of my body, but part of me. Though I had them in my hands now, some day they'd have to go. My husband came into the bathroom, sat silently on the closed toilet, and watched me shake.

I knew that if I talked to my parents, I'd become hysterical, so I texted or e-mailed my father—neither of us can remember which—and he broke the news to my mother, who said it was "the worst day of my life." "It was just awful when you got the news. I just started shouting no, no, no." She went down into the basement to hide her cries from my father and paced and wept and shouted for an hour. "I just couldn't believe it," she tells me more than three years later. "I was so upset. I just felt sick to my stomach and just really felt so awful."

Over the coming months she would revisit her own breast cancer diagnosis from the perspective of her own parents. They'd been out of the country on vacation when she received the news, and she hadn't wanted to worry them so she'd waited for two weeks until they returned to let them know. "I realized how hard it is to see your child go through something so difficult. It was just awful, Lizzie," she tells me.

In BRCA families, these little generational realizations seem to happen a lot.

After her hour in the basement my parents consoled each other, and my mother says, "[we] wanted to talk to you right away just to hear your voice." I don't remember the phone call or what George and I did afterward. But it felt like nothing would ever be all right again.

6 | Watchful Waiting

The day my grandpa died, I went for a mammogram. The death of my father's father hadn't been entirely unexpected. He'd been suffering for ten years from aphasia—memory loss—that affected the language center of his brain. First his keen eye for crosswords went, then his ability to read and remember our names. If he needed a spoon for his ice cream, he'd mime it. By the end, he was hardly able to speak at all, lost in a sea of experience unmoored from language. He had been an engineer, a navy officer, and an intellectually curious man. It was horrible to watch. We suspected this strange aphasia, which wasn't Alzheimer's, had some connection to his work overseeing the unmanned drones that flew around nuclear explosion test sites in New Mexico, which he watched without protective gear. Maybe it was true, or maybe we just needed a romantic story to explain, to justify, to make sense of what unfolded before us. My grandmother cared for him deeply, fulfilling what she saw as her duties as his devoted and loving wife without complaint. She spent the last decade of his life bathing him, shaving him, giving him medication for prostate cancer, lifting him when he fell, and generally administering to his needs as he slipped away from us. Until the very end, he maintained his perpetual good humor—he'd often smooch my grandma, because he loved her. He had occasional moments of clarity—not marked by speech but by a sharpness in his eyes that made you certain he could understand what you were explaining to him. Though he had not been particularly musical in life, as he declined it was one of the few ways we could connect with him, so around the Thanksgiving table we would sing "From the Halls of Montezuma," which he remembered from his stint as an underwater mine disposal expert during WWII. He responded to us in spirited gibberish that told us that even now he wanted to participate in our lives, that he loved us and appreciated our singing.

I learned of my BRCA mutation on a Thursday evening, the day after my grandpa became ill with possible pneumonia. It was clearly the end, so my parents went down to Tennessee that weekend to sit with the other relatives and watch Roscoe, who was Gov to his grandkids, die at home in his bed. I had been invited, of course, but I couldn't go—I didn't think I could bear to watch this up close while still reeling emotionally from the genetic news. Over the weekend, he took a turn for the worse. And so, the following Wednesday morning, I got the call. "He's gone," my stoic father said, voice cracking. I'd only ever seen my father cry once, when we watched *Goodbye, Mr. Chips* as a family in my teens, though my mother swears he cried at my birth too.

That afternoon, out of respect for my grandpa, I put on all black, not knowing that it marked the beginning of a fashion era for me, and jumped into the car to head to my mammogram. I had thought about cancelling but decided against it; I was simply too nervous about my BRCA diagnosis, and this would be my first mammogram in several years. Of course, my grief added to my already-fraught emotions about the procedure, and today, nothing would go well for me. For starters, the woman at the front desk explained to me that they didn't do mammograms on women as young as I was then—twenty-seven. "But I have the gene," I managed to force out. "My mother got cancer at thirty." If the women in my family had waited until the recommended screening age of forty, they'd all be dead. "No exceptions," the receptionist said. She noted that my new insurance company hadn't approved the procedure and pointed me to the phone in the next room. I teetered on the verge of hysteria, could feel myself protesting ever more shrilly, could feel my eyes well up. I would call the insurance company in the next room. I rifled through my purse for my insurance card only to discover that in my wrought emotional state, I'd left the house without my wallet. Of course. Even if I could somehow work this out without my insurance card, I couldn't pay my part of the bill. I stalked back to the car and scrounged in the cushions for enough money to pay for parking—in my haste to leave the office, I hadn't validated my ticket.

So began a frustrating, emotionally taxing experience that often involved arguing for the necessity of mammograms and MRIs despite my young age—what part of MY MOTHER GOT CANCER AT THIRTY did these people not understand?—an insurance preapproval process that was never fully explained to me, and often being treated as a crazy hypochondriac by the people doing the screening. I felt like I was losing my mind—it was so much paperwork and arguing that I could barely focus on the details of what I was supposed to be doing to understand my various scans.

One might imagine a treatment center geared to people like me, where I could simply go in every six months for screening and get all the tests at once. It would be like some sort of rational paradise where I only had to spend two hours a year on this genetic condition. Instead, I had to patch together my own treatment, managed largely by institutions that were not accustomed to BRCA patients and their surveillance needs. I'd even heard horror stories from other women about doctors who weren't as informed about BRCA as they should have been, doctors who told them that a family history of cancer "didn't count" if it was on their father's side, or who were not up to date on the surveillance protocol. And by the way, BRCA women are supposed to have a whole lot of surveillance. It involves a battalion of doctors—by the time I'd been through it once, it seemed like I'd been felt up and fingered by every clinician in the state. According to the National Comprehensive Cancer Network (NCCN), an alliance of twenty-five of the world's leading cancer centers that publishes standard of care guidelines for many sorts of cancer-related treatment, among them, surveillance for BRCA patients, if you want to keep both your breasts and ovaries, then you must scan the ever-loving snot out of them, though to be fair, not even medical professionals have much enthusiasm for ovarian screening. Once or twice a year, a professional should feel you up for lumps. Once a year, your breasts get squeezed tightly between two plates while radiation is shot through them to make a picture—a mammogram. Because many BRCA patients receive screening while they have young, dense breasts that are hard to read via mammography, and because MRIs are a more sensitive scan, you're supposed to alternate mammograms with MRIs. Preferably, each MRI happens on certain days of your cycle and involves having dye pumped into your veins while you lie prone in a giant tunnel with all your body jewelry taken off while the loud clicks of a magnet get pulsed through you. As it turns out, I am one of the unlucky few with an allergy to the contrast dye used, which means I have to take steroids on the day of, or occasionally a pretty intense dose of Benadryl, which addles me.

That sounds like two trips to the doctor: one for a mammogram and one for an MRI, right? Wrong. It's four—two trips to the oncologist who feels me up and orders the tests, and then two to the lab to have different sorts of beams shot into my breasts. Of course, if they find any "suspicious" areas—and the more you scan your breasts the more likely this is—then you have a breast ultrasound, where they squeeze jelly over your boobs and glide a wand over them, using sound waves to look inside. Sometimes this happens at the same time as your other appointments. Worst case scenario is that you have to make a new one.

According to oncologist Dr. Otis Brawley, chief medical and scientific officer of the national health organization the American Cancer Society, part of the reason for this surfeit of appointments lies in how doctors get paid. "I treat patients with cancer," he tells me, "and this is actually a national shame. We see the patient on Monday, [and] write their chemo orders to be administered on Tuesday, because if the chemo is administered the same day we see the patient and write the chemo, we get paid less." Similarly, he says, if I were to go to a high-risk clinic and see multiple doctors on the same day, my insurance would pay as if it had been a single visit—all the doctors would be splitting the reimbursement. It's not efficient, and it's certainly inconvenient for patients, but it's the way the system currently works. My rational paradise does not exist in part due to bureaucratic red tape.

If they do find something inside your breasts during an exam, often they don't tell you at the appointment either. They'll simply say they want to look a little closer and then get back to you days or weeks later. During that time, you wonder whether you screwed up on this round of Russian roulette—the ten-bullet gun has as few as four or as many as eight or nine rounds in it, after all—and whether this time there will be something in the chamber when you pull the trigger.

If breast cancer surveillance in BRCA patients is an exercise in anxiety, then ovarian cancer surveillance is an exercise in futility. Doctors have limited tools—a physical pelvic exam, a transvaginal ultrasound, and a blood test—that have not been proven effective in catching ovarian cancer or in preventing death from the disease. Still, like many other BRCA patients, I go through the motions; surely doing something is better than doing nothing. My gynecologist doesn't seem to think the screening tools are worth it, but I make her screen me anyway. I want that lubed-up wand up my vagina, peeking around for tiny cysts, even though every time I go in the technician meets me with questions—"Why are you even here? What is the reason for this ultrasound?"—leaving me to ponder her apparent judgment of my sanity. Am I crazy for using an imperfect test to screen for a cancer that is rare in the general population? Or is the medical profession gaslighting me? Am I really a hypochondriac, or do this doctor and this technician just not understand the risk I face, why I want my blood drawn even though the substance they test for is only produced by certain types of ovarian cancer and can fluctuate for other reasons, like stress or your period?

NCCN recommends getting into a study on ovarian surveillance in high-risk women, because there is not enough data on this, but my gynecologist is focused on delivering babies, not on catering to whiny women like me. Since my ovarian surveillance comes through her, though, that's another three to four appointments a year—for some reason I can't have my ultrasound during my yearly Pap smear and have to come back for another test. We argue about blood draws too—she thinks they're useless, while according to my research, the efficacy of doing these every six months for high-risk patients has not been evaluated yet. Sometimes I've used up all my arguing points on the biennial ultrasound though, so I succumb to her wishes and skip it. This is one of the horrors of BRCA—the fear that you might know more about BRCA than your doctor and not get the care you need if you follow her advice, balanced against the fear that you're paying her for her expertise, so she probably knows more than you do, and that you might be acting like a crazy, high-maintenance bitch. This is why specialists get paid the big bucks.

Altogether, that's a minimum of seven trips to the doctor each year for screening—sometimes more if you need an ultrasound or a biopsy. This is in addition to any standard maintenance, such as going to the dentist or for a physical or to see specialists like an eye doctor, allergist, and so on. I don't like to spread them out—I try to do them all in one or two weeks. But it is a lot of doctors' visits, accompanied by uncertainty and anxiety around the test results. Watchful waiting has its price. The question is whether it's worth that price, and when I looked into the issue, I found that the answer is surprisingly complex.

While most body tissues—everything from mouth to rectum—can develop cancer, doctors only regularly screen healthy people for a few, among them breast, cervical, colon, rectal, and lung cancer. Currently scientists are studying screening for other cancers, including prostate and skin cancer, according to the National Cancer Institute. We don't screen people for every cancer, just for the common ones that are easy to detect. Just as it doesn't make sense to screen say, both men and women for prostate cancer, because the number-one risk factor for prostate cancer is being a man, it also doesn't make sense to screen everyone in the general population for comparatively rare cancers like kidney cancer, which is hard to detect anyway. On top of this, screening must do more than simply catch cancer—the goal of screening is to reduce cancer deaths and elongate life span by catching cancer early.

But "early" is a relative term when it comes to cancer. Consider two women undergoing screening for breast cancer who have clean mammograms performed on the same day. A year later, at their return appointments, a mammogram discovers a small, slow-growing tumor in one woman's breast. Her cancer is probably curable through traditional therapy. The other woman also has a small tumor in her breast, but the cancer is aggressive and has already hitched a ride through the lymphatic system to her liver and lungs. She has only a few years to live. Although the time between screenings—one year—is identical, the mammogram has caught the first cancer "early" in the development of the tumor and the second one "late" in the development of the cancer. Catching cancer early means catching it when it's in a less lethal stage of the disease, and that depends, of course, on what particular strain of cancer you have. One of the arguments against mammography as a screening tool is that it's good at catching slow-moving cancer that isn't likely to kill you but pretty bad at catching the aggressive stuff that grows fast.

Detection tests for cancer walk a thin line. Ideally, such tests would catch all cases of cancer—they wouldn't return positive results for things that looked like cancer but weren't (false positives leading to overdiagnosis), but they also wouldn't miss any cases of cancer (false negatives leading to underdiagnosis). And while we're wishing for an ideal world, perhaps we could have one test that found all cancers and then a second test to tell you which cancers were lethal and which ones weren't, Brawley suggests. A perfect detection test would be, in the words of Goldilocks, "just right." But since we don't live in a fairy tale, in the real world, over- and underdiagnosis are balanced against one another. You can think of cancer screening like a net strung across a river, fishing for salmon. If I weave it tight enough to catch all the salmon that come downstream, chances are good it'll also catch a bunch of not-salmon—turtles, otter, river weed, stuff that I'm not interested in. On the other hand, if I weave the net loosely enough to let the otter and turtles pass, some of the smaller salmon will swim through—I won't catch them all. Use a very sensitive test, and you'll get lots of false positives. Use a not-sensitive test, and you'll get false negatives. It's pretty hard to win the game.

Further complicating the picture is that it's not enough for a screening test to merely catch cancer—it must also increase life span, and that's tricky to study because of lead-time bias. Siddhartha Mukherjee explains the complications with a thought experiment. Imagine two identical twin women with identical forms of cancer that develop at exactly the same time. The first twin is all about prevention and screening, and after a routine mammogram, doctors discover her cancer in 1995. They treat her with surgery and chemotherapy, but in 2000 she relapses and dies. Her sister dislikes and distrusts cancer surveillance, so she doesn't go to the doctor until 1999, when she discovers a lump in her breast while showering. The doctors give her some palliative treatment, but she dies in 2000 at the same moment as her sister. Did surveillance help? Well, the first twin survived for five years after diagnosis and the second twin survived only one—doctors might say that the first woman's surveillance helped her live longer after diagnosis. Yet the two women had the exact same sort of cancer and the same life span. The surveillance simply pushed the diagnosis date backward, a phenomenon scientists call "lead-time bias." A good detection test, therefore, does not merely make patients survive for longer; it increases overall life span. If the first twin lived longer than her nonscreening sister, maybe then the regular mammograms were worth it. As one can well imagine, it's hugely complicated to produce a good screening tool that balances over- and underdiagnosis and increases life span.

Surveillance rests on the idea that there is a long march from normal cells to atypical cells to cancerous cells and finally to tumors and metastasis. The reasoning goes that if it's found when it's baby cancer—cute and cuddly—but before it becomes Voltron, it's easier to kill. This idea of carcinogenesis, that cancer starts as normal cells which slowly become atypical and then malignant, suggests that catching cancer earlier in its development, when it is not cancer but "pre-cancer," can help prevent cancer death. But not all cancer is suited to these methods. It's a bit like cooking onions—if you put them over high heat, you have to watch them carefully to make sure they don't burn. But if you put them on low, then they take a longer time to go from raw to burnt, and you don't have to be so attentive. Let's say I'm predestined to get cancer. If my cancer is aggressive, the cells in my body might change from normal to atypical to cancerous very quickly, which wouldn't give researchers much time to disrupt this process. If my cancer grows slowly, surveillance is more likely to help, because there's a longer time window in which treatment can stamp it out. Of course, no one knows quite how high their flame is—how fast their potential cancer might grow—in advance.

This is actually one of the cool things about cervical cancer—it tends to grow slowly, over a period of years, beginning on the outside of the cervix and spiraling inward. This long cooking time means medicine has plenty of time to catch it before it becomes lethal. It used to be the leading cause of cancer death in women, but now it's not even in the top ten, thanks to the Pap smear screening test. And we owe the Pap smear to its namesake George Papanicolaou and, oddly, to the fact that guinea pigs don't visibly menstruate.

Greek scientist George Papanicolaou arrived penniless in New York in 1913. Despite his training in medicine and zoology, at first he struggled to get a job, so he worked, I kid you not, as a carpet salesman until he landed a position at Cornell University studying guinea pig menstruation. Guinea pigs had a mysterious cycle, as they didn't bleed or shed much tissue during their monthlies. So Papanicolaou used a nasal speculum to hold them open while he scraped cells off their cervixes and then scrutinized the resulting slides under a microscope; he discovered that the cells waxed and waned in size and shape depending on when in the cycle he'd scraped them. Maybe what was true in guinea pigs was true in humans. By the end of the 1920s, Papanicolaou was studying humans, and as Mukherjee puts it, "His wife Maria, in surely one of the more grisly displays of conjugal fortitude, reportedly allowed herself to be tested by cervical smears every day." As it turned out, human cervical cells also change depending on the stages of a woman's menstrual cycle.

This was interesting but hardly revolutionary. I mean, sure, it's fascinating to know that cervical cells change shape during the menstrual cycle, but no woman needs a pelvic exam and a microscope to tell her Aunt Flo is in town. So rather than focusing on healthy women, Papanicolaou began reading slides from women with many different gynecological problems, from infections to tubal pregnancies and fibroids. In 1928 at a eugenics conference, he announced his shattering discovery. He'd found that smears with abnormal cells in them came from women with abnormal cervixes, and he'd been able to diagnose cervical cancer from such a smear. Over the next decades he refined the procedure, and by 1950 he hit on the idea that its true use wasn't diagnosing cervical cancer, but catching it early, since cervical cancer begins in an exterior lesion before tunneling inward.

As it turns out, the Pap smear is pretty good at catching cancer when it is a fluffy bunny. A 1952 study of 150,000 women in Shelby County, Tennessee, found 555 cases of invasive cervical cancer and 557 cases of preinvasive cancer or precancerous changes. The clincher was that the women with pre-cancer were about twenty years younger than the ones with cancer. The Pap smear meant doctors could detect cancer about twenty years earlier than they'd previously been able to, and it changed cervical cancer from mostly lethal to mostly curable.

While the cure rates were wonderful, the Pap smear also opened the door to uncertainty. It separated the job of the physician—taking the smear—from the job of the pathologist who read it. And as the American Cancer Society promoted widespread screening in 1945, a new problem presented itself—pathologists didn't want to search a haystack of slides for the needle of cancer. A new class of worker—the cytotechnician—was born. Poorly paid and mostly female, cytotechnicians screened slides and brought the hinky ones to pathologists for a second look. The increase in manpower, er, womanpower made widespread screening possible. It also relied on human judgment, which meant it was susceptible to human error. A 1956 experiment that sent twenty smears of atypical cells to twenty-five pathologists revealed how deep human error could run. Three pathologists found no cancer, three found one case of cancer, four found nine cases of cancer, two found twelve cases, and two found thirteen cases. In addition, abnormal pap smears open the door to further surgeries—cervical biopsies to confirm the presence of cancer, or cryosurgery or laser therapy to burn or freeze off any atypical areas. These days, hysterectomy is only occasionally recommended, but back in the 1960s, the common practice in the United States was to treat cervical lesions—or any other lady-organ complaint—with the surgery. As Ilana Löwy, a senior researcher at the French National Institute of Health and Medical Research, wrote, although the Pap smear did reduce deaths from cervical cancer, "the low cost of elimination of cervical lesions is not, however, a zero cost," especially when it comes to precancerous diagnoses. In addition to rare but potentially serious complications, "one of the rarely discussed drawbacks of such screening is an irreversible generation of uncertainty. . . . Diagnosed with a potentially threatening condition with an uncertain meaning, they [the patients] are not sure if they should see themselves as sick or healthy."

As a woman positive for a BRCA mutation, I bear this uncertainty doubly, both because I am frequently screened for cancer and am therefore more likely to receive ambiguous results, but also because the BRCA test itself is a sort of screening for pre-cancer. I may not have any precancerous lesions inside me, but I have been told that I have a potentially life-threatening mutation inside every cell of my body. After my genetic results came back, I no longer felt like the physically healthy twenty-seven-year-old newlywed that I was. Instead I became someone who went to the doctor more than ten times a year, like a good patient, to make sure I wasn't sick yet. I lived in a state of betweenness, in a no-man's-land straddling the worlds of sick and healthy.

The Pap smear's success as a cancer screening tool influenced the later push for mammography, because it proved screening could prevent cancer death and that American women were cool with letting doctors nose around their most intimate parts on a yearly basis in order to do so. It also vindicated the idea that healthy cells change slowly into cancerous ones and that disrupting this progression could prevent cancer. However, the later push for mammography also ignored a bunch of stuff that made cervical cancer a good candidate for screening. Doctors can look directly at the opening of your cervix with their naked eye and a speculum—it doesn't require X-rays or anything—which means screening is easy and not too expensive. Most cervical cancer also moves glacially, like a turtle with a dagger strapped to its head—which means it's easier to catch early. Nonetheless, the same logic behind the Pap smear would be extended, in the 1960s and 1970s, to mammography, and beyond, to "every cancer we had a test on," Brawley tells me.

Berlin surgeon Albert Salomon was one of the first people to take photos of breasts—three thousand breasts amputated in mastectomies, that is—using X-rays in 1913. Sure enough, he'd been able to see cancer and mineral deposits. Unfortunately, his studies couldn't continue because in the mid-1930s the Nazis removed him from his position. He escaped to Amsterdam, leaving behind his technology, which he called mammography. The method didn't resurface in America until the 1950s and 1960s, when the desire for breast cancer screening converged with radiologists who had been honing their craft since the turn of the century.

As with all screening tests, mammography was a nightmare to evaluate, and there are only a handful of large-scale studies that have investigated its efficacy. When it comes to scientific evidence, there's a hierarchy, Brawley says. At the top of the hierarchy are prospective randomized trials—big trials that have two arms, one with patients receiving whatever treatment is being tested, and one with patients who aren't, or who are receiving the current standard of care. So to figure out whether one regimen of new chemo drugs is better than the standard treatment, for example, patients in one arm of the trial receive the new treatment, and those in the other arm receive the current cocktail. In order for the study to be good, the patients should be randomly assigned to one arm or the other, and the rest of their treatment should be identical. Randomized trials are the gold standard of evidence. Next, there are large cohort studies, which follow large numbers of people for long periods of time and examine risk factors. For example, a large cohort study might look at dads who exercise and dads who don't, to see if exercise has any bearing on developing prostate cancer. Below that is a case control study, which compares groups of people with different outcomes—say people with lung cancer and people without—and tries to uncover the differences that might have caused their cancer or lack thereof. And last in line is the opinion of an expert.

There have been prospective randomized trials of mammography—the gold standard—and all eight of them showed that screening saves lives, according to Brawley. Unfortunately, "because you're dealing with human beings and because you're dealing with large numbers of human beings, no study is ever going to be as precise as we would like it to be," he tells me. Most of the studies have potentially large flaws, according to Mukherjee. A 1963 study of eighty thousand New York women aged forty to sixty showed that mammography reduced breast cancer mortality by 40 percent, but he writes, it was later criticized because the control group hadn't been told it was part of a trial, which had skewed the numbers. As Brawley tells me, the number of women enrolled in the trial changed over time, making it hard to measure the prospective benefit. Unfortunately for future studies, the New York results prompted the American Cancer Society to launch its massive Breast Cancer Detection Demonstration Project (BCDDP), which enrolled 250,000 women nationwide in 1971. With so many women already participating in this project, it was hard to find enough uninvolved women to run a good study in the United States. So scientists tried in five different Swedish cities, Scotland, and Canada. Pretty much all of them had problems, according to Mukherjee.

The Canadian trial in the early 1980s suffered because nurses randomized patients after taking down their medical histories and steered a disproportionate number of high-risk women and women with abnormal exams into the mammography group, muddling the results. Mukherjee called the Edinburgh study "a disaster" because "doctors assigned blocks of women to the screening or control groups based on arbitrary criteria. Or, worse still, women assigned themselves. Randomization protocols were disrupted. Women often switched between one group and the other as the trial proceeded, paralyzing and confounding any meaningful interpretation of the study as a whole." Only the Malmö trial of the late 1970s bore any fruit. It studied forty-two thousand women, screening half, and by 1988 the results were in. Mammography had benefited women over fifty-five, reducing breast cancer deaths by 20 percent. Young women had not benefited. As Mukherjee puts it, "This pattern—a clearly discernible benefit for older women, and a barely detectable benefit in younger women—would be confirmed in the scores of studies that followed Malmö." Of course, years later, a paper in the *Journal of the National Cancer Institute* suggested that the Malmö numbers were compromised because breast cancer rates among Swedish women may have been spontaneously declining prior to the introduction of the mammogram.

Screening women in their forties for cancer via mammography is still controversial. As Brawley tells me, cancer shows up white on mammograms, and many women in their forties still have dense breasts, which also show up white. Looking for a white tumor on a white background is much harder than looking for a white tumor in a less dense, older breast that shows up grey or black on the scan. On top of this, cancer risk goes up as you age—there aren't so many women in their forties who develop breast cancer, so you have to sift through more of them to see a benefit. Brawley tells me that the data show that you'd have to scan 1,904 women in their forties to save one life. And in saving that one life, numerous other women would undergo callbacks and unnecessary biopsies. Mammography saves lives, he says, but for women in their forties it takes a lot of work to save one life, and in the meantime, the data show that some women will have such a bad experience with the scares of screening that they will swear off mammography for the rest of their lives—including in their fifties and sixties when it is much more effective. These ambiguities around risk and benefit make the wisdom of screening women at age forty unclear.

What Brawley says about screenings scaring some women really resonates for me. If BRCA women are told to scan their breasts even more, I wonder whether that makes us—a high-risk group that may be more frightened of cancer after potentially watching family members die—even more likely to avoid the emotional ordeal of screening as well. The numbers show that of the BRCA women who decide on surveillance, most of them are vigilant for the first few years, but by five years later the vast majority of them no longer undergo the recommended regimen.

Adding to the confusion is that the more we look for cancer, the more we seem to find it. A study that examined the breasts of women who died in unrelated accidents found that mammography had only discovered 18 percent of the tumors identified by pathologists who looked at many thin sections of each breast. Mammograms don't catch everything, and in fact, that's good. As Löwy puts it,

> An enhancement of radiologists' capacity to see tiny clusters of microcalcifications does not automatically produce a more desirable outcome for the patient. Excessively sensitive image-generating machines may identify more precancer and thus produce more clinically ambivalent results. And vice versa: less sensitive radiology equipment can produce a lower proportion of false positive results and detect less "pseudo-disease" (that is, changes in tissue that will not produce clinical symptoms of cancer in a woman's lifetime).

Sensitive detection tests may pick up pre-cancer that would not necessarily become lethal and kill you, as in the clinically ambiguous diagnosis of LCIS, lobular carcinoma in situ. LCIS is a condition predictive of invasive cancer, according to Dr. Mary Daly, who chairs the NCCN Guidelines panel for genetic/familial high risk assessment for breast and ovarian cancers. Women with LCIS have a higher likelihood of developing invasive cancer later, but not necessarily in the same spot or even the same breast as the original LCIS. The condition is not cancer but rather a marker of increased risk for cancer. Still, once you know you've got this predictive condition, will you be able to rest without doing anything? In this way, detection tests can lead to medical interventions, such as surgery or chemoprevention, which may feel psychologically necessary while not necessarily prolonging life.

In fact, there's good evidence that we're overtreating breast cancer, Brawley tells me. "We can tell from breast cancer in the United States that 20 percent of the breast cancer we diagnose was not going to kill anybody if it was not diagnosed. . . . Twenty percent looks malignant but is programmed to behave benignly." Some flavors of breast cancer will metastasize and kill you, while others will just hang out in your breast and do nothing. Unfortunately, we don't know how to tell them apart—they both look like cancer under a microscope. And so women with cancer are subjected to arduous treatments. Still, there's an upside. As Brawley puts it, "Yes, we cured some people of breast cancer who didn't need to be cured. But my prospective randomized studies tell me that we cured some people who needed to be cured [too]." The problem is even worse with prostate cancer, where it's 60 percent of localized tumors that don't need to be cured, Brawley says.

When I began looking into mammography, I thought I would find easy and clear evidence that it definitely helped reduce cancer death. Instead, I feel more confused than ever. Malcolm Gladwell posed one of the core questions about mammography in the *New Yorker* in 2004. "You'd expect that if we've been catching fifty thousand early-stage cancers every year, we should be seeing a corresponding decrease in the number of late-stage invasive cancers. It's not clear whether we have," a fair point, although according to the National Cancer Insitute's most recent statistics, the breast cancer death rate declined on average by about 1.9 percent per year between 2001 and 2010. If screening works, how come up to 30 percent of all breast cancers are still destined to become metastatic killers? A 2011 Danish meta-study of mammography trials found that "reported reductions in breast cancer mortality cannot be explained by differences in screening effectiveness" and that "screening appeared ineffective." And a 2012 paper in the *New England Journal of Medicine* that examined mortality data for women over forty found that mammography only reduced the diagnosis of late-stage cancer by about eight cases per one hundred thousand women. As the authors concluded, "Despite substantial increases in the number of cases of early-stage breast cancer detected, screening mammography has only marginally reduced the rate at which women present with advanced cancer." Is potentially saving eight lives worth giving one hundred thousand women mammograms, which carry with them the risk of further exploratory surgery, such as biopsies? It depends, I selfishly suppose, on whether I am one of those eight.

When I ask Dr. Brawley about the Danish study and about Gladwell, he tells me that they both "have their point that metastatic disease has stayed at a stable rate for the last forty years. The people who are pro screening have a point that there's a 30 percent risk reduction in breast cancer mortality risk." It's unclear why breast cancer death has decreased, but Brawley suspects it's probably a combination of advances in cancer treatment, advances in screening, and improvements in physical health. "The one thing we've proved reduces breast cancer risk is physical exercise," he reminds me.

So where does that leave the young BRCA woman? In a pretty confusing place. Mammograms may or may not be effective. Breast MRIs, which use magnetic waves to look inside women's breasts, penetrate the breasts of young women more easily but also have a higher rate of false positives than mammograms. As Brawley explains it, "If I do an MRI on ten women age forty-five that I pull off the street, I'm going to find an abnormality in half of them, and with those numbers, probably none of the abnormalities are going to be cancer." And that's not all. "The other catch-22 is that we don't know that finding these lesions through MRI also saves lives. Many of us, myself included, believe that early detection is better than late detection, but we have to separate what we believe from what is scientifically proven," Brawley says. On one level, BRCA carriers like me are at a high risk of cancer, so statistically it makes more sense to screen us than women in the general population. Or as Brawley puts it, "We want to do MRI on high-risk women because the MRI so commonly finds something that looks like cancer, that if we only image the people who are at high risk, the likelihood that the mass is high risk is much better."

Still, so far, adding MRIs to mammograms mostly suggests that you're three to five times more likely to get called back about false positive results and that there's a pretty good chance your cancer will be detected—but it says nothing about whether such screening actually does the job of preventing cancer death.

On top of all this, screening for breast cancer involves shooting radiation into your breasts to take a picture via mammography. And guess what can also cause cancer? Radiation. Of course, radiation is all around us—we are exposed to about 3 mSv (millisieverts) of radiation each year from stuff like radon gas at home, cosmic rays, and so on. A mammogram exposes women to about 0.4 mSv of radiation—about the same amount as what you absorb in seven weeks of living. That's not very much. Still, radiation exposure is cumulative. It's one thing to get a yearly mammogram once you turn fifty. If I live to eighty, that's thirty years of mammograms and about 12 mSv of radiation—the same amount I'd absorb in four years of living. Although now NCCN recommends delaying mammography in BRCA carriers until age thirty, at the time I began screening, physicians recommended that I start screening ten years before my mother developed breast cancer—that is, at age twenty. Assuming I live to eighty, that would be sixty years of annual or biannual mammograms, and at least 24 mSv of radiation—the same amount as eight years of living. As the American Cancer Society website puts it, "Most studies on radiation and cancer risk have looked at people exposed to very high doses of radiation, such as uranium miners and atomic bomb survivors. The risk from low level radiation exposure is not easy to calculate from these studies." How paranoid is it to wonder whether my intensive screening might begin to contradict its primary purpose?

Not all that paranoid, as it turns out. "Quite honestly, there's a lot of discussion in the professional community," Dr. Brawley tells me, particularly because BRCA is active in repairing DNA, and radiation from mammography damages DNA, raising "the theoretical possibility that women with BRCA are more likely to get breast cancer [from mammography] than women without BRCA," and a few studies support that claim. Dr. Daly is more reassuring. She reminds me that as technology has improved, the doses of radiation used in mammography have decreased and that much of the historical data on radiation revolves around people who had major chest radiation due to treatments for Hodgkin's or who had stuff like acne treated with radiation in their teens. "So it's not clear any of that historical data applies to current practice," she says.

The bottom line, Dr. Brawley tells me, is that it's complicated. "Cancer screening is always harmful and sometimes beneficial," he says, "And we need to figure out when the screening test has benefits that are greater than the harms." The benefit-to-harm ratio is a complicated calculus. Lung cancer provides a good illustration of the point, because we know a lot about it, in part because people who contract lung cancer tend to die within a few years, which means it's easier and cheaper to study—in contrast, because women often live decades after developing breast cancer, studies require more participants and longer, costlier follow-up. In the 1960s, the ACS did a study on lung cancer screening using X-rays, Brawley tells me. In 1975, the results were in: the screened arm of the trial had a higher death rate than the unscreened arm because screening found suspicious zones, which led to surgical biopsies that ended in death. "It was death by biopsy," Brawley says. Even now, he says, we know that lung spiral CT, a scan that uses radiation to look at soft tissue, "actually does save lives. However, for every 5.4 lives saved, there is one life that is lost to medical complications." Or consider breast self-exams, the favorite rah-rah cause of each October. Two studies performed in Chinese and Russian women—more than 388,000 of them—showed that setting aside thirty to forty-five minutes each month to really feel your breasts thoroughly had zero impact on breast cancer mortality. The only thing it did was make women more likely to find suspicious zones that might lead to more biopsies. Women being aware of their breasts—touching and watching them in the normal course of business—is what is being promoted now. Finally, Brawley tells me about an old study on ovarian cancer done in a few thousand nuns that estimated that screening one hundred thousand nuns would uncover three cases of localized ovarian cancer. "You would end up saving three women from ovarian cancer and killing five women from surgery to figure out why they had an abnormal screen," he says. The harm-to-benefit ratio is no good. More surveillance is not necessarily better.

So where does that leave someone at high risk, someone like me? Most of the studies done have looked at screening in the general population and not in women at high risk of developing breast cancer. Truly, I'd like to see some studies performed on my risk group to determine the basic efficacy of ovarian and breast surveillance in preventing cancer death. I shouldn't hold my breath, though. Large studies are expensive to produce, weighed against the comparatively small number of BRCA carriers who would benefit, and they require lots of patients. Are there even enough BRCA women who would be willing to be randomly assigned to the screening or nonscreening arm of such a trial? Would such a study even be ethical? I guess living with the flawed human screening we have now, and living with the uncertainty generated, is all part of what it means to be a mutant.

7 | A Tale of Too Many Mastectomies

If we had to pick one, I'm unsure whether my family would select an Amazon or Saint Agatha for our mascot. The Amazons, a mythological tribe of warrior women, cut off their right breasts so they could better draw their bows, exchanging femininity for fearsomeness on the field of battle. They're even named for their breastlessness—a common derivation suggests it comes from the Greek *a* (without) *mazos* (breast). At the other end of the spectrum, there's Saint Agatha, a Sicilian virgin who dedicated herself to God in the mid–third century. The Roman prefect Quintianus cut off her breasts because she wouldn't have sex with him or sacrifice to the Roman gods. You can tell that mostly male artists painted her, because she's often shown holding a plate with her disembodied breasts on it, gazing over them with dead eyes and a smooth, untroubled expression, as if these body parts meant no more to her than a plate of bread. In fact, she is the patron saint of bread and bell makers because her severed anatomy resembled buns and bells. She's the patron saint of breast cancer patients too.

The Amazons and Saint Agatha represent two extremes in how we narrativize cancer patients. On one hand, they are fierce, defiant warriors taking control of their health and doing the difficult thing to improve survival, no matter the side effects. On the other hand, they are martyrs, submitting to vicious disfigurement at the hands of the medical establishment, giving up something valuable to protect what is even more precious.

But people are complex and contain multitudes. Cancer patients or BRCA patients are neither simply martyrs nor warriors. In some form, both narratives imbue each mastectomy. Patients are actors in their own fate, and acted upon by modern medicine. With this in mind, I'd like to look at two sorts of mastectomies, separated by untold medical advances, an ocean, and more than a century of time. The first, performed on the novelist Frances Burney in 1811, treated her cancer. The second ones, performed on my mother's cousin Kathy in 1979, and Lisa, in 1983, were done to lower their risk of cancer and occurred thanks to the empowerment of the women's liberation movement of the 1970s. Ironically, between 1811 and 1979, the treatments for cancer and cancer risk took opposite paths. Surgery for actual cancer became less draconian, shrinking from removal of the entire breast to simple removal of the tumor. As scientists learned to pinpoint cancer risk with increasing accuracy, treatment for high-risk patients took the opposite path, culminating in the prophylactic mastectomy.

The novelist Frances Burney's account of her own mastectomy—one of the few patient-written narratives passed down from history—lays bare the underlying brutality of the operation. Her description in a letter to a friend nine months later is one of the most visceral, horrifying, death-metal things I've ever read. But if she could endure it, we should be able to read about it, no? (If you are squeamish, you may wish to jump to p. 144.)

In 1810, she developed a lump in her breast. Cancer. Napoleon's surgeon, Baron Dominique Jean Larrey, vowed to operate. She didn't want to spend months dreading her operation, so she asked the surgeons to give her only a few hours' notice. One morning, a letter came. The surgeons would arrive at ten o'clock that day—in two hours. So she tried to prepare herself, even though she "had to disguise my sensations and intentions" from Mr. Frances Burney, who was not keen on the procedure, and whom Fanny wanted to protect from watching her suffer. She had to get him out of the house, so she asked her son Alex to contact a colleague of her husband's and to tell him to make the supposed business urgent and important enough to keep her husband out of the house the whole day. "Speechless & appalled, off went Alex, &, as I have since heard, was forced to sit down & sob in executing his commission," she wrote.

Then she told the doctors she needed a few more hours—until 1:00 PM—to get ready. She fixed up a room for her husband to stay in while she recovered. And with that in order, she was ready to face the dread procedure—but one of the attending doctors was running late and couldn't show up until 3:00 PM. The suspense was awful. "This, indeed, was a dreadful interval," she wrote. "I had no longer anything to do—I had only to think—TWO Hours thus spent seemed never-ending." I feel you, Frances. She stumbled into the salon, where the sight of the bandages, compresses, sponges, and other surgical materials made her feel a little sick.

In case she died during the operation, she wrote notes to her husband and son. Then she drank a single wine cordial—I hope she's underestimating her consumption for propriety's sake, because if I knew what was coming next I'd get drunk off my gourd—and seven men in black entered her room without even knocking. As it turns out, this emotional nightmare—the contemplation of her own death, the dreadful anticipation of the operation, the tears of her son—would take a distant second to the physical nightmare that was about to begin.

They asked her to mount the bed in the middle of the living room but she couldn't move, locked in a moment of horror. "I stood suspended, for a moment, whether I should not abruptly escape—I looked at the door, the windows—I felt desperate." But of course, there is no escape from cancer, except the purifying fire of surgery. After a moment, "my reason then took the command, & my fears & feelings struggled vainly against it." Her maid wept by the door, and her two nurses stood "transfixed." The doctors tried to send the women away, but Fanny resisted them. "No, I cried, let them stay!" Two of the women broke and ran off, but one defiantly remained.

The women weren't the only ones afraid. The doctors stammered at Frances. One of them tore paper into tiny bits; another had a face "pale as ashes."

And screw anesthesia. Anesthesia is for pansies—or for operations conducted at least thirty years later. They placed an ordinary cambric handkerchief over her face and called it a day. It was quite thin; she still could see the squadron of doctors and "the glitter of polished Steel" through it. They uncovered her breast, and one of them made a circle in the air with his finger, indicating that they would take the whole thing off. This freaked her out, so she ripped the handkerchief off her face and sat up, explaining that all her pain radiated from a single point in her breast. But the doctors told her again that it must all come off, and firmly put the cloth back over her face.

Then they started sawing off her breast while she watched them through the handkerchief. Here's how that felt:

> When the dreadful steel was plunged into the breast—cutting through veins—arteries—flesh—nerves—I needed no injunctions not to restrain my cries. I began a scream that lasted unintermittingly during the whole time of the incision—& I almost marvel that it rings not in my Ears still! so excruciating was the agony. When the wound was made, & the instrument was withdrawn,

> the pain seemed undiminished, for the air that suddenly rushed into those delicate parts felt like a mass of minute but sharp & forked poniards, that were tearing the edges of the wound—but when again I felt the instrument—describing a curve—cutting against the grain, if I may so say, while the flesh resisted in a manner so forcible as to oppose & tire the hand of the operator, who was forced to change from the right to the left—then, indeed, I thought I must have expired.

From then on, she kept her eyes shut so hard "that the Eyelids seemed indented into the Cheeks." For a moment, she thought they were done, but the cutting resumed. "Dr. Larry rested but his own hand, &—Oh Heaven!—I then felt the Knife [rack]ling against the breast bone—scraping it!—This performed, while I yet remained in utterly speechless torture." She passed out at least twice from pain during the whole thing and could not speak of the operation for months. It's no wonder that even a single question about her ordeal "disordered" her. "Even now," she wrote, "9 months after it is over, I have a head ache from going on with the account! & this miserable account, which I began 3 Months ago, at least, I dare not revise, nor read, the recollection is still so painful." The operation lasted only twenty minutes, "a time, for sufferings so acute, that was hardly supportable."

The procedure was hell on her husband, who must have found out about the operation as it was happening. He added a few lines to her letter: "No language could convey what I felt in the deadly course of those seven hours." Reading Fanny's letter affected him. "I must own, to you, that these details which were, till just now, quite unknown to me, have almost killed me, & I am only able to thank God that this more than half Angel has had the sublime courage to deny herself the comfort I might have offered her, to spare me, not the sharing of her excruciating pains, that was impossible, but the witnessing so terrific a scene, & perhaps the remorse to have rendered it more tragic. for I don't flatter myself I could have got through it—I must confess it."

Fanny allowed seven men to do this to her, and she faced the horrific ordeal with unimaginable courage. She's a martyr and an amazon. The miracle of this surgery, in an era without antibiotics, is that she survived the operation and didn't die from postsurgical infection. And her cancer—if indeed it was cancer and not a benign lump—did not recur.

Frances Burney's story, the story of how we used to treat cancer, is one of surgical radicalism. Until the advent of chemotherapy and radiation therapy in the twentieth century, surgery represented medicine's main tool for treating breast tumors. The Edwin Smith Papyrus, which proclaimed breast cancer as having no cure, also recorded the cauterization of an identifiable breast tumor with an awful-sounding tool called a fire drill. According to a description written in 1296, Leonidas of Alexandria dealt with breast tumors by using a knife to remove the breast, and then cauterizing the wound. The ancient physician Galen was picky about which tumors he'd carve out—they had to be easily accessible. He also cut widely around the tumors to ensure he removed all of the mass—a practice that ensured clean margins—and he eschewed burning surgical sites because it damaged surrounding tissue. In eighteenth-century Europe, surgeons used horrible devices—bladed rings or pairs of blades—to cut off breasts swiftly, a procedure that often led to hemorrhage and disfigurement. From the late 1700s through the end of the following century, science sped along at a fast clip, although advances were patchy at best—known in certain areas of the world, but not others.

Japanese surgeon Seishu Hanaoka, for example, developed and experimented with anesthesia on his wife and his mother more than forty years before the West started investigating the field. In 1805 he put a woman under and performed a mastectomy—quite possibly the world's first painless breast removal—and by the end of his career he'd performed 150 of them. Unfortunately for the rest of the world, Japan lived under its *sakoku* policy of isolation at the time, which prevented the spread of medical breakthroughs. Meanwhile in France, surgeon Jean-Louis Petit published a work on mastectomy that recommended the removal of breast, lymph nodes, fat, and part of the pectoral muscle—a Halsted before Halsted.

All of these developments, combined with advances in anesthesia and the discovery that surgery required sterile conditions, set the stage for the radical operations of both William Halsted and William Meyer, published separately in 1894. Both operations removed breasts, lymph nodes, and different sections of pectoral muscles but detached items in a different order. Halsted's meticulous method took four hours, while Meyer slashed time by using scissors for blunt incisions.

The brutal procedure became the preeminent breast cancer treatment for the next half century. It saved women's lives but left them stoop-shouldered and with limited arm mobility. The Halsted mastectomy is a relic of its era, when surgery was often meant to be palliative, not curative. Doctors saw plenty of late-stage tumors—many likened them to various sizes of bird's eggs—cases probably so far advanced they'd be incurable even by today's standards. Rather than let tumors ulcerate and burst through the skin, giving patients miserable, pained demises, doctors whacked off breasts as a humanitarian effort. That Halsted's mastectomy was able to improve the relapse rate to a mere 52 percent was extraordinary. Given the sorts of cancer he saw, it's understandable that he believed it to be a local disease with a local cure.

The upswing of less drastic procedures, such as lumpectomy during the twentieth century, is not merely the story of scientific advancements, such as chemotherapy and radiation. It's also the story of women rising up and demanding breast-conserving therapy; it's the story of how women's lib upended the relationship between mostly male doctors and their female patients.

In our culture, breasts have significance in a way that, say, the spleen does not. They are visible and eroticized markers of femininity, so breast surgery necessarily engages with our ideas of gender. As cancer culture historian Ellen Leopold points out, historically, the patriarchy has structured the relationship between mostly female patients and mostly male surgeons. She writes:

> At its most reductive, the aura surrounding breast surgery reinforced the worst gender stereotypes, attributing all power to a male hero and all frailty to a damsel in distress. The surgeon was alert, erect, and skilled, and the patient, asleep, supine, and helpless—that is, without animating or humanizing virtues of any kind. Life-saving surgery, in other words, seemed to require the total degradation of a woman's spirit as well as of her flesh. This abasement, so integral to the surgical ordeal, was to color every aspect of treatment for most of a century.

Many early- to mid-twentieth-century practices for treating and talking about breast cancer prove Leopold's point. As she discusses, in the 1960s the authority of mostly male surgeons was unassailable. Surgeons could also decide whether to tell patients the truth about their conditions—doctors too had trouble choking out the word "cancer." One survey from 1961 found that 90 percent of physicians did not tell patients the truth—that they had been diagnosed with cancer—preferring euphemisms like "mass," "lesion," and "tumor." Doctors themselves viewed cancer with hopelessness and did not like delivering tough news; they also wanted to shield patients from feeling hopeless after a diagnosis. Interestingly, while doctors overwhelmingly preferred not to tell patients about their cancer diagnosis, patients—89 percent of them—overwhelmingly favored knowing their own diagnosis, according to a 1950 study. This inequality in knowledge affected breast cancer treatment. As Leopold points out, "The fact that her surgeon was unable to communicate the results to her [the patient] directly did not deter him from acting on them unilaterally, that is, without her agreement." In the early 1950s, breast cancer survivor and advocate Fanny Rosenow called the *New York Times* to post a notice for a breast cancer support group she was starting. Her call ended up routed to the *Times'* social editor, who greeted her request with a pregnant silence. "I'm sorry, Ms. Rosenow, but the *Times* cannot publish the word *breast* or the word *cancer.* Perhaps you could say there will be a meeting about diseases of the chest wall." She hung up, disgusted, but through the persistence of Rosenow and her fellow survivor Teresa Lasser, the organization Reach to Recovery was founded, a program my grandmother later participated in as a cancer survivor, visiting other patients in their hospital rooms to talk about what to expect. Public discussion of breast cancer in obituaries, for example, also garbed breast and ovarian cancer deaths in euphemisms such as "women's cancer," or "prolonged illness." The culture seemed embarrassed about cancer in general, but about breast cancer and other women's cancers in particular—it silenced women, denying their experiences to each other, refusing even to tell them the name of the malady killing them. The heroic Halsteds of the world rode in on their white horses and carved out internal organs. It was viewed for so many years as the only way. Even after chemotherapy and radiation became available, until the mid-1970s it was standard practice to put a woman out for a breast biopsy and then remove her entire breast in one go if the tumor tested positive to avoid the inconvenience of putting her under general anesthesia twice. These breast biopsies mirrored the larger struggle around women's rights in the 1960s and 1970s. Who should have dominion over women's bodies—the women themselves or the male doctors who thought they knew best?

So let us sing the praises of journalist, breast cancer activist, and cancer patient Rose Kushner, who went after the one-step biopsy practice in the mid-1970s. In 1974 she developed cancer, and she wanted some time between biopsy and breast removal to decide on a course of action. She had to visit nineteen surgeons before she found one willing to biopsy her tumor but not remove the breast. Eventually, she testified before NIH on the matter, arguing that a two-step procedure separating mastectomy from testing would allow surgeons to better assign stages to cancer and offered women the option of making up their own damn minds. After all, it wasn't the surgeon's life at stake but his patient's. Now, the two-step biopsy procedure is the worldwide standard for treatment. As it turns out, Rose was a friend of my cousin Kathy, who participated in her uprising, attending basement meetings for her new patients' right groups in the early 1980s. Thanks to the advocacy of Rose and other women in the movement, my mother had a day to figure out what sort of treatment she wanted—a huge psychological improvement over my grandmother, who suffered great trauma from waking up after a biopsy with a brutal Halsted mastectomy.

The same year that Rose Kushner developed breast cancer, so did Betty Ford, who brought breast cancer out of the closet by candidly discussing her diagnosis and mastectomy with the press. She was not suffering from a "woman's cancer." She had breast cancer. B-r-e-a-s-t cancer. She spoke the unspeakable words in public, giving interviews to reporters and letting them photograph her in the hospital, and that changed everything. As a 1987 *New York Times* piece—written thirteen years after Ford's diagnosis put it, when Ford got cancer, "many Americans still considered mastectomy a taboo subject, too fearful or even shameful to be discussed openly." Within a few weeks of Ford's diagnosis, the *New York Times* reported a four- to tenfold increase in women requesting screening at the American

Cancer Society's twenty-seven free centers. According to her 2011 *Time* magazine obituary, the former First Lady's diagnosis set off a screening tidal wave so large that "the reported incidence of the disease rose; some researchers even called this the Betty Ford blip."

Forty years later, in 2013, Angelina Jolie would do the same thing for women with BRCA mutations by writing about her preventive mastectomy in a *New York Times* editorial. Her piece and the ensuing media coverage did several things for BRCA carriers. For starters, BRCA mutation carriers no longer have to preface conversations about surgery with a lecture on genetics and risk. Now, women can simply say, "That thing Angelina did? I did it too," and the average dinner-party attendee will understand. Second, having a cancer-causing BRCA mutation is pretty complicated, and the breast-removal option sounds draconian until you start thinking through the bigger issues. A lot of people have called BRCA women who elect to have mastectomies crazy. That a big-name celebrity made this same decision in consultation with, presumably, some of the finest doctors money can buy, is a powerful vote in its favor. And although the long-term effects of her announcement are not yet known, a *Washington Post* article published a few months after her op-ed found that DC-area genetic counselors were fielding twice as many calls about genetic testing. Joy Larsen Haidle, president-elect of the National Society of Genetic Counselors and a practicing counselor herself, told me, "There has definitely been a Jolie effect." She explained that immediately following the announcement, she and most of her colleagues "received a dramatic increase in calls asking about the breast cancer genes . . . and for many of us that trend stayed elevated for several months afterward." A Harris Interactive poll performed in August of that year found that 86 percent of women had heard about Angelina's mastectomy, with 5 percent saying they would seek advice about their breast and ovarian health.

Unfortunately, the idea that women might not be smart enough to make their own medical decisions is still around. After Jolie's announcement, the media did publish some hysterical, reactionary articles that insulted women's intelligence—will more women get unnecessary mastectomies after Angelina Jolie? It's a major medical procedure, not a manicure. If Idris Elba had his testicles removed for a really good reason, do you think men would jump at the chance to be just like him? How stupid do they think we are? Some of the articles confused different concepts of preventive mastectomy, conflating the removal of the remaining healthy breasts in women who had already had cancer (contralateral prophylactic mastectomy) with the removal of both healthy breasts in high-risk women with BRCA mutations, and with contralateral mastectomy in women with BRCA mutations who had previously suffered breast cancer. What science we have shows that prophylactic mastectomy in BRCA women prevents breast cancer. To compare this group to the general population of cancer survivors is misleading and inaccurate.

Still, like Betty Ford and breast cancer patients, Jolie has brought BRCA patients out of the closet. And back in the 1970s, as it became more acceptable to talk about breast cancer, the women's lib movement also made strides in allowing women to assert sovereignty over their own bodies—for example, the right to abortion guaranteed by *Roe v. Wade* in 1973. As women demanded control over themselves, the relationship between doctor and patient changed; doctors no longer held a position of unassailable authority over patients. Rose Kushner and fellow journalist Betty Rollin published pieces questioning the necessity of radical mastectomies and heckled surgeons at medical conferences about how radical surgery had never been properly tested in a controlled environment.

In 1971, the Halsted mastectomy had its eightieth birthday, the anniversary of Halsted's first description of the procedure. And that year represented the beginning of the end for that operation. Spurred on by patient activism, surgeon Bernard Fisher (known for the wonderful saying "In God we trust. All others [must] have data.") began studying lumpectomy and radiation. He spent ten years gathering data and found that lumpectomy—a breast-conserving operation in which the tumor and some surrounding tissue are excised—plus radiation worked just as well as radical mastectomy. His research represented the formal fruition of the studies that Geoffrey Keynes had first embarked upon after World War I.

However unwelcome they are, tumors come with a degree of certainty. If you have a tumor in your breast, doctors will probably try to cut it out, and if the cancer has spread to your lymph nodes, those might go too. The problem has a location, a physical manifestation that may be pinpointed. A tumor presents a clear problem; your risk for cancer is no longer vague. It's actualized, solidified, and therefore terrible, as terrible as the methods of treatment—surgery, radiation, hormone therapy, and chemotherapy—treatments a blitz of scientific studies has evaluated.

Take away the tumor—take away the cancer—and only the vagaries of risk remain.

For much of the twentieth century, medicine put a whole host of women into the "at high risk for breast cancer" category, prescribing mastectomies despite the fact that their effectiveness in preventing cancer had not been scientifically studied. That's not too surprising, considering Western medicine's troubling history of removing lady parts with only slight justification.

Since surgery became tenable after the introduction of antisepsis and anesthesia in the second half of the 1800s, the medical version of the track coach's "Walk it off" has been "Surgically remove your uterus and/or ovaries." Doctors prescribed oophorectomy and/or hysterectomy for a wide variety of physical and psychological conditions. Irregular period? Epileptic? Nymphomaniac? Precancerous lesions detected on your cervix? Let a historically male surgeon just nip in there and remove those pesky lady parts. After all, if you're not going to get pregnant, they're useless, right?

Of course, these organs aren't useless. As we now know, the removal of ovaries has been linked to osteoporosis, hip fractures, dementia, short-term memory loss, loss of libido, cardiovascular disease, and more, to say nothing of how the loss of fertility or removal of internal organs might impact a woman's self-image. Removing the uterus through hysterectomy—the second most common operation (after cesarean section) undertaken by a staggering one in three women in the United States—carries fewer risks, other than the complications like blood clots associated with any surgery and loss of fertility. Removing healthy organs has serious consequences.

Prophylactic mastectomy has more than a century-long history that traces the reverse path of breast cancer surgery. Even as enthusiasm for mastectomies on cancer patients waned, the idea of removing breasts to prevent cancer in healthy women gained traction. The history of preventive mastectomy wrangles with technological advances that permitted better cosmetic outcomes and made women more willing to undergo such surgery. It also raises questions of risk and risk tolerance: How high must a woman's risk of breast cancer be before doctors recommend such surgery? Is desperate fear of breast cancer sufficient? Or must she have premalignant changes in the breast as well? And what does pre-cancer even look like?

While breast cancer surgery began with radical mastectomy and went toward lumpectomy, preventive breast surgery began with lumpectomy and advanced toward mastectomy. An 1882 paper by

New York surgeon T. Gaillard Thomas on the removal of noncancerous breast tumors without "mutilation of the organ" describes good candidates for the procedure, along with the medical and psychological reasons for the operation. Tumors should be medium sized—in Thomas's time, that meant about the size of a duck's egg, though such tumors would be considered huge today. Medically, he wrote, the procedure carried only slight risks and could prevent a smaller tumor from growing dangerously and degenerating from "a benign into a malignant growth." His untested assumption—that benign lumps can turn into cancerous ones—loomed large in the future of prophylactic mastectomy. He also said that benign tumors could make women really nervous, an anxiety that surgery could ease. Many women with BRCA mutations would find his description of anxious female patients familiar: "I have found that the mere presence of a tumor in the breast usually concentrates upon it the thoughts and attention of the patient, impairs her happiness, and renders her apprehensive, nervous, and often gloomy." Yes, that sounds familiar, though I might have used a stronger word than "gloomy." A woman with a tumor, he wrote, spends "a great deal of her time" checking out the lump "and in comparing notes and asking information concerning it of her female acquaintances, and the result of all this is very frequently to engender a state of mental disquietude and wretchedness for the relief of which a resort to extirpation of the tumor is an entirely defensible procedure." That's true—only now, thanks to a century of technological advances, we don't just call female friends; we give ourselves hypochondria by reading too much medical information on the Internet. The lumpectomy of giant benign breast tumors, Thomas argued, had medical and psychological justification. Because such operations were preventive, not curative, they had to cross a higher bar of doing all this without mutilating the patient.

Thirty-five years later, St. Louis surgeon Willard Bartlett proposed an operation that fit the nonmutilating bill. Like Thomas, he agreed that prophylactic mastectomy had psychological motivations. In a paper advocating mastectomies for women with lumpy breasts, he claimed that "the breast is of such psychic importance to the female patient" that women waited until too late to consult surgeons. "It is the fear of having the breast mutilated that keeps patients away and allows a tumor to run a progressive course," he wrote. Again, he talked about waiting too long, his words pitting feminine vanity—the psychic importance of breasts—against the curative knowledge of the surgeon. The proposed solution, he thought, was to give women new breasts to satisfy "the psychic element." To this end, he described a mastectomy very different from Halsted's. Rather than remove the entire breast, skin and all, Bartlett removed the breast tissue through an incision made at the bottom of the breast where it met the torso—the inframammary fold—and preserved the skin pocket. After cauterizing the interior, he replaced the scooped-out breast tissue with fat moved from the woman's thigh or butt. Though the details of the surgical techniques have changed and improved over time, the core idea is correct. This sort of operation is considered one of the exemplary standards of breast reconstruction today—a subcutaneous (skin-sparing) mastectomy followed with flap surgery to put your own tissue into your breasts.

Bartlett didn't seek to remedy cancer but rather pre-cancer through this operation. He recommended the procedure for women with "chronic fibrocystic mastitis [lumpy breasts], which is generally admitted to be a pre-cancerous condition." The idea that cystic mastitis—a condition that affects more than half of women at some point in their lives—will inevitably turn into cancer was still a large untested leap, but one that located the vagaries of breast cancer risk physically and palpably in benign tumors. His assumption that lumpy breasts mean cancer risk, and that risk can be carved out of the body, lingered for many decades.

The faulty belief that lumpy breasts increased a woman's risk of breast cancer persisted for several reasons. Surgeon Joseph Bloodgood—one of Halsted's pupils—provided some of the reasoning in his 1931 paper on borderline breast lesions. Cancer and lumpy breasts feel pretty much the same to the hand of the doctor, for example, and reasonable pathologists, examining slides of breast tumors under microscopes, often disagree about whether lumps are benign or cancerous. Rather than attributing this uncertainty to the imprecise science of pathology, which relies on human judgment as well as scientific guidelines, Bloodgood focused on the fact that benign but atypical cells and cancerous cells can look similar. From this he concluded that mastitis and other breast changes "must be looked upon as border-line breast lesions." This ambiguity—mastitis isn't cancer but may develop into it—put the surgeon in a rough spot. He had to decide between two crappy alternatives: "The danger of an incomplete operation for cancer and the unnecessary mutilation or removal of the breast for innocent lesions." Today, of course, thanks to women's lib, it's patients who have to make this wrenching decision, not doctors. Faced with uncertainty in 1931, though, Bloodgood thought it was safer for the patient to have the full mastectomy.

Interestingly, the advance of science caused this headache, Bloodgood argued. Because women knew more about breast lumps—knew that cancer was no longer an automatic death sentence—they began to visit doctors earlier and earlier in their illnesses, so doctors began seeing cancer when it was young and tender and eminently killable. Maybe mastitis was simply baby pre-cancer.

Preventive skin-sparing mastectomies provided better cosmetic outcomes, but uncertainty about who should have such operations dogged surgeons for decades. By the 1940s, mastectomy for cystic mastitis was on the wane, though when confronted with indefinite biopsy results, many surgeons still opted for the "safer" route of mastectomy. The creation of the silicone implant in the 1960s, which enabled even better reconstruction, drove prophylactic mastectomy to greater popularity, though due to differences in terminology and the patchy way medical advancements become adopted, it's hard to know how many preventive skin-sparing procedures were done. A mid-1960s survey found that nearly half of all plastic surgeons had performed prophylactic operations, though a 1969 survey found that such procedures accounted for only a small percentage of breast operations.

Prophylactic mastectomy in the 1970s was marked by more uncertainty than a back-room dice game—uncertainty about who should have the procedure, what constituted breast cancer risk, what mastectomies worked best, and even whether mastectomy worked at all to prevent cancer. Throughout the decade, a debate raged between plastic surgeons about whether and when the procedure was justified. Sure, the operation spared women the physical pain of multiple biopsies and the mental pain of fearing cancer, but did it actually prevent the disease? It'd be more than three decades before the science began to come in. In the meantime, surgeons performed prophylactic mastectomies on women for a whole host of conditions—benign tumors, cystic mastitis, breast pain, fear of cancer, history of biopsies or cancer scares, family history of cancer, and desire to keep breast shape.

In the meantime, Bloodgood's observation that the education of women caused doctors to witness more and more ambiguous strains of breast cancer came into full fruition with the widespread introduction of mammography as a screening practice in the 1970s. Greater sensitivity in screening tests doesn't necessarily correspond to better clinical results. The X-rays picked up stuff that doctors couldn't feel with their hands, and once they'd discovered a lump, they felt driven to biopsy it just in case. Mammograms can also detect micro-calcifications—tiny deposits of minerals scattered throughout the breast that can indicate nascent breast cancer, stage 0 cancer, pre-cancer that is not quite cancer but not quite not-cancer either—and lead to diagnoses like lobular carcinoma in situ (LCIS), clinically ambiguous results that indicate an elevated risk of developing life-threatening disease.

As mammography became standard in the 1970s, large numbers of women without symptoms underwent screening, which led to a rise in diagnosis of these conditions. Should such breast cancer risk—localized to the cellular level—be treated with mastectomy or not? At the same time, the movement against mastectomy and toward localized surgical solutions was gaining traction, and women demanded more control over medical treatment. This shifted the burden of decision making from surgeons to patients. As French medical researcher Ilana Löwy put it in her book about women and preventive surgery, "In the 1970s, women diagnosed with breast cancer increasingly insisted on their right to participate in decisions about their treatment. The proposal to burden women with the responsibility of making a wrong decision when facing medical uncertainty might have been an unanticipated consequence of this demand."

On top of this, it was unclear what sort of mastectomy worked best. A simple mastectomy removed the entire breast en masse but made reconstruction more difficult. Subcutaneous mastectomy, in which surgeons scooped out breast tissue but left the skin, provided better cosmetic results. However, those results depended on having a nice, thick, even flap of skin to cover an implant, which meant balancing cosmetic outcome against the medical result—leaving some breast tissue might improve the flap but compromise the breast-tissue-removing purpose of the operation.

Finally, there wasn't much data on whether prophylactic mastectomy worked. Some women experienced complications related to their implants, and a few went on to develop cancer. Plastic surgeon Vincent R. Pennisi, a tireless advocate for prophylactic skin-sparing mastectomy, did a few studies on the procedure. His 1974 survey of 460 plastic surgeons who performed 4,179 such operations over a span of ten to fifteen years found that only twenty-four patients developed breast cancer—about 0.5 percent, and a drastic improvement over the national rate of breast cancer diagnosis. But this survey had problems—surgeons self-selected into it, and it did not have a control group, rendering the data difficult to interpret. Pennisi's follow-up study five years later found that only two women out of one thousand followed for nine years developed cancer, but this study suffered from similar problems.

It's hard to get good data on prophylactic mastectomy. For starters, it doesn't lend itself to a control group, or to blinding the study from patients or physicians. Given that I'm at high risk for breast cancer, would I really let some researcher randomly assign me to have a mastectomy or keep my breasts? Hell no. The vagaries of risk make forming studies difficult as well. In order to examine whether mastectomy reduces cancer risk, scientists would have to know the starting risk of the women they were studying. And in the 1970s, that wasn't easy—did cystic mastitis really put women at higher risk for cancer? How much did heredity play into it? The identification of BRCA1 in 1994 would bring these issues into sharper focus, but until then, it was the Wild West.

In this uncertain era of medicine, my great-aunt Elfrieda urged her daughter Kathy to explore prophylactic mastectomy. At the time, El had terminal ovarian cancer and was busy dying. Suddenly, Kathy says, she'd show up to visit her mother at the hospital, and El would have research on prophylactic mastectomy there. Since she was being treated at the hospital where she'd taught nursing, she had many friends around. Once, when Kathy visited, "all of a sudden there are these two guys in her room," her mom's gynecological surgeon and his plastic surgeon colleague who did reconstructive surgery and just "happened to be there." Kathy knew her mother had arranged for this "accidental" interaction. The surgeons came armed with pictures of the new reconstructive procedure and supposedly wanted Kathy's opinion because they were about to start teaching the new methods. "I guess she [El] was doing that because she wanted me not to avoid it," Kathy says. "She was also telling me, 'It's different now, they don't do the Halsted, you can have reconstruction.' " While El slipped in and out of comas, Kathy quietly met with her mother's doctors and obtained referrals to internists and surgeons. She never talked to her mother about it, though. "Since she was suffering so much that last year, I did not want the focus to be on me."

El's final days were grim. The cancer had spread to her spine, and she spent her last eighteen months in the hospital or sitting on the couch at home. Lisa says that for her mother, "it was a great effort to do anything at all." Her father, Ralph, taught El about football and how to watch football so they could have something to do together. Lisa was in college and had planned to spend a semester abroad in England. Before she left, El told her, "If anything happens to me, I want you to stay and have a good time." She didn't want her kids to miss out on anything because of her illness.

El dealt with her ovarian cancer with black humor, which meant "there was nothing we couldn't ask her," Kathy says. For the last two years she wore a wig and navigated in a wheelchair. The daughters watched and helped her. It would take her two hours to get up and get ready, and then she would go out to visit friends or relatives, resting for forty-five minutes after she arrived, visiting for an hour, and then resting for another forty-five minutes to get the strength to go home and go to bed. "But she would do it because she needed to feel like she was a real person and needed to leave the house," Kathy says. At home, Ralph gave her the garage door opener to press when she needed him if he was out doing yard work. El's sickness had ups and downs. "It just became normal for her to be very, very ill and then to have a little rebound," Lisa says. "Every time the phone rang, I wondered whether this was the call."

The pain was so excruciating that at one point—as the story goes—El asked one of her sons for marijuana in hopes it might help. As Lisa puts it, "It's what cancer does to families. They just sort of break down. Other things don't matter anymore when someone is that ill." As she died, El charged Kathy with two tasks: to make sure Lisa paid attention to the family legacy as soon as she turned twenty-one, and to find a holiday on which the family would get together so it would survive after her death. Historically, women have provided the social glue that holds a family together—when many women in a family become ill and die, well, families can fall apart. With Lisa, the youngest, halfway through her senior year of college, El's death came at a transitional moment. "We were at the age when we were trying to be individual," Kathy says. "It had huge implications for the family, and in some ways we never recovered. Each of us dealt with it in our own way. But the family unit never recovered, and of course that affects my kids."

The extended family had broken too. My grandmother Meg had always been competitive with her sister, and she also suffered from periodic bouts of depression and psychosis. Relations between the two sisters had been strained, and at the time El suffered so horribly, my grandmother was dealing with her own drama—a second diagnosis of breast cancer. "Because of the rift, and because of what Meg was going through," Kathy says, "I had no one." She couldn't talk to her father, in part because he was depressed, but also because, Kathy says, "I needed a woman." Lisa was away at college, and her mom's best friend was so devastated that she burst into tears while trying to comfort Kathy more than once.

Amid all of this, and spurred on by her mother's encouragement, Kathy decided to have a mastectomy for a number of reasons. She'd had her first needle biopsy for something the doctors told her could be in situ cancer at age twenty-one. Since then, she'd had biyearly mammograms and physical breast exams every three months. At age twenty-seven, she'd already had four needle biopsies, and given her family history, the doctors recommended multiple yearly screenings. "The idea of going to have my boobs checked out four times a year—I was afraid I'd become a hobbyist. It felt like it was going to take over my life," Kathy says. She also had dense breasts that made telling lumps from biopsy scar tissue and normal tissue difficult. "Part of it was just that if I kept having biopsies at this level, I would destroy my skin so much that it might not be possible to have reconstructive surgery," she tells me. Then, of course, there was the fear—the fear that what had happened to her aunts Trudy and Meg and to her dying mother could happen to her. Due to all the cancer in the family, Kathy felt ambivalent about her breasts. "They didn't seem fun," she says. "They seemed threatening." Finally, the way her mother had dealt with the loss of her breasts made Kathy believe she'd be able to bear the pain. Growing up, she says, her mother radiated this attitude of "look at me in my marriage; I am beautiful and feminine, and you will never have the surgery that I have. Life is more important than beauty."

Her father offered his total support. "He was there for me," Kathy says. "He basically said whatever he could do." But she felt her husband from what she now refers to as her "rehearsal marriage" was absent while her mother was ill, and he thought her plan to have a mastectomy "was crazy and a reaction to my mother dying and that I would regret it the rest of my life." My mother remembers hearing about Kathy's plan—she, too, found it extreme at the time, though three years later, after her own breast cancer, she would urge her own sister to have the same procedure.

Something beyond Kathy pushed her on. "I think there was another fear too. I was afraid that if I didn't follow through with it, I would never do it. It must be how my son felt when he bungee jumped. There was part of that as well. Like now, even just talking about it now."

Twelve male surgeons told Kathy she was having a hysterical reaction to her mother's illness. "They said no good doctor removes healthy tissue," Kathy says. The thirteenth agreed to operate.

Now Kathy calls that time "the most intense period of my life." To cope with her mother's imminent death, her impending mastectomy, and her failing first marriage, Kathy threw herself into work and wrote a book.

In January of 1979, El died. Hundreds of people came to the funeral, Kathy says, "shocking numbers, people frantically getting folding chairs to extend. And a scholarship was set up for her. But she died young and tragically [in] a small town."

Soon afterward, El's original breast surgeon and friend, now retired, had Kathy and Lisa over to his house, a cool modern affair. Lisa remembers, "We're sitting there on his couch and he's telling us that we ought to have this surgery, and I remember it seemed so outrageous to me at the time. I couldn't even get my mind around it. It sounded so horrific and dramatic, and he was so insistent. He said with the family history and the age at which these women got cancer . . ." Lisa couldn't begin to relate to the conversation. "I'm so out of here," she thought. But her sister, of course, was already deep into the planning stages.

Three months after El's death, Kathy and a friend visited the hospital the day before her scheduled operation to do the pre-admitting stuff. While she filled out paperwork, an administrator walked into the room and told her the surgery had been cancelled. Kathy felt devastated. After everything she'd been through—a marriage on the rocks, her mother's death, the emotions around voluntarily cutting out her breasts, and meetings with thirteen surgeons—this was one barrier too many. It was too much, too much to ask of a person, to do this. If she didn't do it now, she wasn't sure she could do it at all. They simply could not cancel on her. She broke down and wept in the office. The administration used the excuse that they hadn't gotten pre-approval from her insurance company, apparently, even though that "was never done in those days," she says. They wanted her to pay for everything up front. Sitting in that office, crying, Kathy wondered how she could pay for it all. Maybe her father could help. Eventually, she got ahold of her plastic surgeon. They couldn't cancel it. They wouldn't cancel, he said. He stormed into the administration office fresh out of surgery and still in his scrubs. And then, as Kathy tells me thirty years later, her voice rough and on the verge of breaking at the memory, "The plastic surgeon pulled out his credit card and paid for it. The whole thing. And at the time, oh my God, it was thousands of dollars. It was a six-hour surgery with two surgeons, and it was thousands of dollars." When Kathy's father heard, he drove several hours to Chicago to hand the man a check, shake his hand, and thank him.

The surgeons wrote up Kathy's operation for the insurance as exploratory surgery of her breasts due to previous biopsies. She had a subcutaneous mastectomy, kept her nipples, and had silicone implants installed all in a single operation.

Four years later, my mother developed breast cancer, and the Muehleisen curse had trickled down into the second generation, an insatiable dark beast demanding appeasement from my matrilineal line under penalty of death. Although Lisa had resisted surgery after watching Kathy's, now she felt compelled. She remembered the trip to the house of her mother's surgeon, she says. "I could hear the doctor's voice in my head saying that I was a little young at twenty-two. It just seemed like that was enough reason to do it. I don't even remember waffling. In my head, I just had to figure out that medical insurance part of it. My husband at the time, his family just thought I was insane. He thought I was nuts. I went forward with it." At the time, Lisa was about twenty-seven, and, like Kathy, her surgeons went to bat for her, informing her insurance company that if they didn't pay for the operation now, they'd be shelling out for cancer treatment in ten years. The lack of support from her first husband and his family ultimately caused a marital rift so deep that, eighteen months after the surgery, they separated and a year later they divorced. She thinks they would have split up eventually, anyway. The mastectomy "was the catalyst," Lisa says. "The marriage was broken anyway, but it was the trigger that set things in motion." It had been the same for her sister Kathy: the surgery—and her husband's lack of support then and when her mother was dying—set her divorce in motion too.

Lisa's and Kathy's mastectomies share many similarities with Frances Burney's. They all did it without the support of their husbands. They all dreaded the procedure. And for Kathy and Fanny, administrative delays made the agonizing waiting time a little worse. Finally, the weight of uncertainty about the procedure settled over all of them, despite 170 years of medical advances. For Fanny, the uncertainty lay in whether she'd survive at all, caught between her tumor and often-lethal surgery performed in an era before antibiotics. For Lisa and Kathy, the uncertainty lay in defining their risk. The family history appeared to give them a very high chance of developing breast cancer, but the identification of the gene and data proving that prophylactic surgery could successfully lower breast cancer risk were decades away, though we now know, thanks to at least five studies, that mastectomy lowers the risk of breast cancer in BRCA patients by about 90 to 95 percent, and maybe even more. They all made grueling decisions—they chose mastectomy—and for that they are amazons. Yet they are subjects, too, martyrs to the imperfect knowledge of science.

Still, it's queer that as surgeries for actual cancer shrank in scope during the twentieth century, from the Halsted mastectomy to lumpectomy, drastic surgery for something far less palpable—cancer risk—became part of the protocol. And even now, when the most advanced technology of all can look inside your DNA and find the tiniest error in code, the smallest blip of a mistake inside the cell nucleus, it seems ironic that the main treatment offered is removal of an entire organ. But how can one fight something as nebulous as uncertainty? Who can see the wind? How does one fight a ghost?

8 | The Black Cloud

In some ways, my grandpa's funeral revived me. It gave me permission to cry, mostly for him, but for myself too. And it contextualized my genetic results—this wasn't a death sentence, but simply a statistic of cancer risk. An up to 87 percent chance of developing breast cancer in my lifetime. A 40 to 60 percent chance of ovarian. After the funeral ended, my aunt put the box of ashes in my arms to carry home for my grandmother. It felt heavy for a man who shrank to almost nothing in his final days, and when my curiosity forced me to open the polished wooden lid, all I could see was a smaller plastic box. It fit perfectly inside.

I drove from Tennessee back to DC with my parents, and my mother drove me further, back up to New Jersey, where she would stay with George and me for a few days.

Since my mother doesn't sit still, we don't when she visits. We visited the Ikea vortex to look at furniture, different grocery stores for all the cooking we would do, the needlepoint shop in Princeton for a project I wanted to start, and the park, where we walked around and around before dinner.

Our conversation circled too. We shuttled between talking about cancer and carefully not-talking about it. I was still in a state of shock, uncertainty, and terror. I couldn't climb out of the thick of these emotions the way I would have liked to, because what I imagined on my mother's face was potentially guilt, as well as dread that our family's curse might be seeking its next victim. I wished I was in shape to comfort her. She told me about her surgeries and how much she loved me and how glad she felt, after cancer, to have the luxury of dull mediocre days stretching in front of her. She assembled delightful salads with Asian pears and toasted pumpkin seeds. We ate ice cream together, and she taught me how to needlepoint; but eventually she had to drive home.

After she left, George and I tried to return to normal. But the truth was that after the genetic test, the old normal no longer existed, and the new one haunted us.

The gene consumed me, at first slowly, but as the shock wore off it devoured every waking moment not forcibly occupied with other activity. I was OK three mornings a week, from 5:00 to 10:00 AM, when I was summarizing news articles. And I was OK when I worked on my first book, which would find a publisher in only a few months. But I couldn't sleep. I lay in bed and thought about the inevitability of this mutation, my risk written into the DNA in every cell of my body. I knew it upset my husband when I cried, so I tried not to sob unless it was really bad. Whenever he realized I was weeping, he rolled over to hold me, but surely, by this time, he was tired of the grating endurance of this pain. My life operated by rote. In the mornings, I woke like an amnesiac; everything felt blank until the memory of my situation descended in a fresh shock, and I was crushed to resurface into this reality once again. Occasionally I cried while George was in the shower.

During the day, in between bouts of work, I cruised the online forums and scoured the web for other BRCA narratives. They made me feel like less of an alien. I e-mailed friends and researched medical treatments. Too often, my Internet meanderings ended with wallowing in my own misery, thinking obsessively about what could happen to me—the hair loss from chemo, the radiation therapy, how sad George would feel after my lovely funeral—and about how horrible cancer must have been for my mother and grandmother. I developed increased sensitivity toward anything vaguely intended to be moving, from Internet videos on bullying to calculated commercial efforts, like a corny BP campaign about renewable energy. As a kid, I always found it weird that my mom and grandma cried during the cheesy parts of movies, and now I wondered whether this, too, was hereditary.

The sea of tears I navigated daily became its own source of guilt. I was weak willed, I thought. After all, I was still a healthy twenty-seven-year-old newlywed. And I didn't even have cancer.

My uncle Alan remembers this period in his relationship, living under what he called "the black cloud" before my aunt Cris decided what to do about her cancer risk after it spiked as a result of my mother's diagnosis. "She talked a lot about it," Alan recalls. "She's very communicative, and we have a talky kind of relationship. Oh yes, she talked a lot to me about it. And she was frightened. She was frightened for her sister; she was frightened for herself, and when she's frightened she talks." Cris found the topic inescapable. As Alan tells me, "She knew the best thing she could do was not think about it and not worry about it because there was nothing she could do about it. . . . When it affected her most she would even have long periods of weeping and even attacks of anxiety, and as I said, other times she was fine. It was not a constant pall over our lives, but it was definitely part of it."

My father remembers living under that same cloud, one that my mother's breast cancer diagnosis made particularly black and threatening. My parents didn't live with the uncertainty of whether my mom would get cancer; they dealt with uncertainty about whether she would survive. What he and my mother went through after her diagnosis sounds like hell. As he tells me, "I thought that [her dying] was a real possibility. And of course she was real worried about it, you know, because of the family history. She was worried she wouldn't get to see you grow up and that sort of thing. And of course there's just the emotional aspect of seeing the physical consequence of the mastectomy, which is pretty shocking. So, you know, you have your image of yourself as a healthy couple, right? And then suddenly you're not a healthy couple." The pall of uncertainty changed more than just my mother's body—it changed his image of their relationship.

Women may experience the primary trauma of a BRCA mutation—in the form of cancer, fear of cancer, mastectomy, and so on—but their partners, who have a lifelong stake in their well-being, also suffer. When I appeared on the *Today Show* years later to talk about my life under the pall of cancer, my grandpa Roy, Ace to his grandkids, watched and put his head into his hands and cried. He'd seen this gene affect his in-laws, his wife, his daughter, and now his granddaughter.

Mostly, George and I dealt with the grief through food, and I craved the comfort fare of my childhood. The day of the fateful phone call, I demanded the easy solace of Kraft macaroni and cheese and Dove bars. The next day, he made sure I got out of the house by sending me to the farmer's market. We had pesto and grilled squash. Then I wanted braised Chinese short ribs with noodles and homemade kimchi and plain tacos in hard shells from a mix, like I remembered my mother making, and though we exchanged some half-hearted banter about how soft tacos with home-braised pork would taste superior, in the end he appeased me. I made watermelon granita—a favorite of my mother's—and because I knew I would eat as many brownies as we made, I dug up an Internet recipe for a single-serving version microwaved in a mug.

We drank martinis, bourbon manhattans, scotch and soda, beer, wine, and occasionally champagne cocktails. I made rosemary and thyme simple syrups and we concocted new elixirs. When we drank, I felt guilty. After all, alcohol consumption has been connected to increased risk of breast cancer. But remaining in my natural mental state for twenty-four hours a day was intolerable. I needed relief. My uncle Alan remembered this too. "She and I have always liked wine," he says, but when the black cloud hovered over his relationship, "every time she had a glass of wine or any alcoholic drink she would be very worried because there were indications this would increase her odds of getting breast cancer."

We turned to television for balm too. We needed to escape to worlds where there was still hope and magic was possible, entertainment with repeating units of reassuring structure, where Xena outwitted her enemy and Veronica Mars caught the criminal and Hercule Poirot explained what really happened, where Dale Cooper loved him some coffee and Buffy slayed the demons of teenage existence and the lawyers argued hard on *Law & Order*. We hadn't had real TV in years so we watched on DVD or our computer, our old favorites soothing us, although through familiarity they'd lost their ability to distract my brain from the ever-present focus. And even this genre television could trigger me. We skipped the *Buffy* episode with no soundtrack, the one where her mother dies—not from vampires but from brain cancer that metastasizes. We tried to pick up *Battlestar Galactica*, but its dour plotline—will the human race survive?—and the breast-cancer stricken president didn't offer the relief we craved. Even Dr. House sometimes lost patients. Neither booze nor fatty food nor television could wipe away the fear that seeped into me the moment my brain was unoccupied.

Over time, George and I began to talk about how I would handle this diagnosis medically. I recognized immediately that at some point—ideally far in the future—my breasts would probably come off, but in the meantime, I really wanted to keep them. In fact, I'd do anything to keep them. We visited the oncologist together and learned in detail about the options.

I had four. I could do nothing, which the oncologist didn't recommend and which seemed impossible to me anyway. I could undergo intensive surveillance—a barrage of tests and scans of my breasts and ovaries every six months in hopes that any cancer that might develop would be detected early, when it was more curable. This method wasn't invasive and sounded quite attractive. I could remove my breasts or my ovaries, or ideally both at some point. If I removed my ovaries, my breast cancer risk would be halved thanks to the reduction of estrogen from my body, and my ovarian cancer risk would drop even more dramatically. Of course, I was only twenty-seven, and oophorectomy causes surgical menopause and infertility. Removing my breasts would slice my breast cancer risk but leave my ovaries intact—with their ability to allow me to bear children, or to explode with cancer. And then, there was chemoprevention.

"What's that?" I asked my oncologist.

It meant taking estrogen blockers such as tamoxifen. Essentially, lots of tissues in the body—including the breasts—have receptors for estrogen. The hormone fits into these receptors like a key into a lock. Chemoprevention drugs put gum into that keyhole, preventing the body from absorbing the hormone.

In a way, the connection between estrogen and breast cancer has been recognized for centuries. Once known as "nun's disease," breast cancer seemed to afflict the habit-clad at a noticeably higher rate. A professor of medicine writing in Italy in 1713 who wondered about the cause of the phenomenon eventually settled on celibacy—an obvious lifestyle difference between nuns and other women. As we now know, being sexually active doesn't raise or lower one's breast cancer risk, but having kids does. Women who give birth—particularly at young ages—and who breastfeed their children for a long time have a lower incidence of breast cancer than women who do not. It's thought that pregnancy and breastfeeding reduce cancer risk by disrupting the menstrual cycle, which reduces total lifetime exposure to estrogen. For the same reason, women who have fewer periods during their lifetimes—because they start menstruating late or have more than one child—also have a smaller lifetime risk of developing breast cancer. Pregnancy reduces ovarian cancer risk too. Still, it's complicated. Pregnancy and breastfeeding don't reduce all breast cancer risk, merely the risk of the more common flavor, which is estrogen-receptor positive. And having kids only helps if you do it before thirty—if you get pregnant afterward, your risk of cancer is slightly higher than that of a childless woman. While having kids lowers risk over a lifetime, in the short term—for about ten years—it slightly raises risk.

Tamoxifen and raloxifene lessen that risk by preventing the body from absorbing estrogen that may nurture incipient cancer cells. Tamoxifen reduces the risk of developing future cancers in women with ductal carcinoma in situ (DCIS), a sort of stage 0 cancer that raises some of the same clinical ambiguity as a BRCA diagnosis. A 1998 study of more than thirteen thousand women at high risk for breast cancer found that tamoxifen significantly reduced the rate of cancer, but that study did not focus on BRCA patients. A subsequent smaller trial of tamoxifen among BRCA patients showed a benefit for BRCA2 carriers, but not BRCA1 carriers, though the sample size was too small to be statistically significant. This may be because tamoxifen reduces the risk of estrogen receptor–positive breast cancers, while estrogen receptor–negative cancers are more common in BRCA1 women. Of course, taking estrogen blockers sounded unappealing. They induce temporary reversible menopause in all its hot-flashing, libido-killing glory, and if you take them for more than a few years they increase your risk of uterine and endrometrial cancers, as well as other rare but potentially serious complications. Basically, my oncologist explained, they could buy me a little time to make a further decision about surgery while immediately lowering my risk.

Also, it was good that I had gone on the patch for a few years in grad school, apparently, because a little hormonal birth control reduces a woman's ovarian cancer risk substantially.

We talked about timing as well. "Are you planning on having children?" the doctor asked, a question that would seem invasive coming from a parent.

"Maybe," was all I could answer. I'd always assumed I would have kids; as a collector of experiences, I was interested in this archetypally human one. But we felt unprepared and uncertain.

If I got knocked up soon, in the next year or two, the doctor said, we could delay other treatments for a while. It's pretty unusual to develop cancer while pregnant, he told us, but it does happen very, very occasionally.

It was a lot to take in within thirty minutes, and I was glad I had brought George to be a second set of ears. I also wanted to force him to confront this diagnosis emotionally and viscerally, to hear the cold facts from a medical professional as I had had to. He had been verbally supportive but distant, almost reserved, as my feelings took a front seat in our relationship. I wanted to push his emotions forward to ensure they were as strong as mine. I wanted to see him grieve with me.

At home, we discussed the options. I discarded chemoprevention first, because it was a stopgap solution. I didn't care if it was only temporary; I didn't want to go through menopause in my twenties, especially because I planned to expedite it anyway. Ovarian cancer is incredibly scary, and I still intend to have my ovaries removed, as the doctors recommend, "after childbearing is completed"—a small, simple phrase for such a big decision—or before age forty. I will have plenty of time to enjoy libido-killing, irreversible, surgically induced menopause when I'm supposed to be at my sexual peak.

The other options were agonizing to sort through. I was twenty-seven, the age at which it's likely that Trudy began developing breast cancer. Maybe. I wore her ring on my finger and thought about how she was too frightened to pull the trigger on mastectomy. If I removed my breasts, I wouldn't be able to breast-feed, and while I wasn't certain I believed that the children were *my* future, the idea of giving up potential breast-feeding—my insatiable writer's curiosity wondered how it would feel—gave me pause. Mastectomy is not reversible. I pictured knives cutting into my flesh and shivered. What about having a baby right now, in the next year, and putting off the decision?

At that time, I worked part-time for a news site and wrote my first book on the side. George was a PhD student in biophysics with another three or four years of school left. We could barely afford our one-bedroom apartment, which was too small for all our books and cooking equipment and old tchotchkes we were too sentimental to trash. There was no room for a baby there. Still, the idea of delaying permanent solutions for as long as possible was attractive to me. George and I went back and forth, round and round on this issue, one we hadn't planned to address for at least a few more years.

Twenty-six years earlier, my mother's cancer had changed my parents' plans for a family. We have trouble talking about this, my parents and I, because they want me to feel loved; they want me to feel like enough. But the truth is that they wanted more children. At the time, my mother tells me, the doctors thought that having pregnancy hormones floating around her body might increase her risk of a cancer recurrence, and they advised her not to have more kids. I always thought that meant maybe one more. Decades later, my uncle explains that this news was "a huge hit to the stomach for them. I think they were hoping to have four or five kids, and this was part of their marital dream, and this put an end to that dream. I had the sense that maybe for Dick [my father]—he wanted to have a flock. And it just meant that this part of the expectation of their life, this was just not going to happen." It was a loss for my father, her illness. As he tells me years later, in addition to his fear for my mother's life, "I remember feeling sorry for myself. Just that, you know, it kind of disrupted my plans for myself. I wanted to have more children, so that was sort of up in the air, and um, I think at that time the doctors were uncertain of what the prognosis was, you know, so there was a good likelihood that things wouldn't turn out as well as they did."

George and I decided that the idea of breast-feeding wasn't a good reason to have a baby before we were ready. I scoured the Internet for articles on the benefits of breast-feeding, on moms who hate on other moms because they don't do it. I began to feel preemptively guilty for my unborn theoretical children and angry at the imaginary horde of moms who might judge me. I concocted vicious retorts to their rude nosiness.

Another meeting with my oncologist confirmed the decision. "What happens in the worst-case scenario?" I asked him. "Like, if I get cancer while I'm pregnant?"

He explained that if it were early in the pregnancy, they would ask me to terminate. If it were late in the pregnancy, he would meet with my ob-gyn and schedule a C-section to remove the baby as soon as it was viable outside the uterus. But really, he said, it was a very small chance.

My mother remembers a similar meeting as she recovered from cancer. She wanted to explore having more children. "My gynecologist said, 'OK, but I want you to sign a document before you get pregnant about what we should do if you get cancer when pregnant. Abort the baby? Have chemotherapy while pregnant? Deliver the baby?' I thought I had a responsibility to stay as healthy as I could for you. I just decided that it wasn't worth the risk." My parents have loved me deeply every day of my life, and this can't have been an easy decision for them. The knowledge has exerted a subtle pressure on me as well, to live up to their expectations, to be better than myself, to fill their lives the way a flock of children couldn't.

Whether the chance of cancer while pregnant was small or not, the choice of aborting a wanted pregnancy or playing chicken with disease until the baby was big enough to live without me sounded awful. That wasn't a position I'd choose, no matter how small the odds.

And so, with a heavy heart, and after numerous calls to the insurance company to make sure the radiology department would actually let me in for a mammogram, I began surveillance again.

At the front desk of the clinic, we played the same scene over. Every time I remember it, it's like I'm right back in that moment, smelling the weird medical ozone mixed with terror. I'm here for a mammogram. They don't give mammograms to women in their twenties. "What part of 'I have a BRCA1 mutation and my mom got cancer at thirty don't you assholes understand?' " I'd said in my head. I kept my smile even and explained that I had pre-approval from my insurance company and I was on the schedule.

This clinic had the trappings of my mother's mid-1990s health club—teal walls, varnished wood trim, TVs for you to watch from overstuffed chairs while you're waiting, little changing rooms with slatted benches in them—but somehow, these did little to conceal the fear that had seeped into the walls. The soothing colors and smiling staff provided only a thin veneer over the reality of maybe-cancer, a veneer I tried to mirror on my face, even though panic had welled up inside me.

I was escorted to one of the changing rooms and given a sanitized, plastic-wrapped gown in a pinky-mauve color. Pink. Why does it always have to be pink? Why couldn't we get something stylish, like cerulean, or zebra stripes, or retro navy polka dots to represent breast cancer? Why did this breast cancer clinic have to remind me—as if I could forget—that I am feminine and my girly female lady woman parts could be sprouting tumors. I loathe the cultural expectation that women adore pink. In preschool, I remember obstinately selecting blue frosting for my class birthday cake, even though I knew the teacher expected me to make a gender-appropriate selection. Pink is a straightjacket of assumptions about who I am and what I like based on the faulty premise that all women are the same. To me, pink represents the shackles of socially constructed femininity, the expectation that I will wax my bikini line and laugh stupidly at simplistic jokes to please the men who tell them. It is supposed to be reassuring, I suspect, like the furniture, a gentle color to paper over the horror that cells inside your body might be uncontrollably redividing. I resent having it shoved down my throat at every opportunity. Pink is a cancer metastasizing into every arena remotely related to this issue. Worried about your breast health? Here, have pink ribbons, pink sweatshirts, pink-wrapped perfume, pink-painted jet planes, pink cosmetics and bras—even the romaine lettuce I buy in the supermarket comes wrapped in pink. Pink pink pink pink pink. It chokes me, especially in October; it never lets me forget for one instant that cancer is waiting to strike, waiting to rot us from the inside. Cancer isn't cute. And the color that represents it shouldn't be the same one used to market Barbies, for gauzy tutus fluttering around ballerinas, for princess getups and bubblegum.

Clearly, I couldn't go out there naked, though, so I put the damn thing on and took myself to the waiting station to watch TV. The mammogram—like all mammograms—wasn't particularly pleasant. Stand here. Hold these handles and stay still. Yes, it is uncomfortable, isn't it, the nurse in colorful scrubs said as she flattened my breasts between the plates. She apologized for putting me through this.

Afterward, I was escorted back to the waiting station. They had found something in my right breast they wanted to examine more closely, the nurse told me, so I needed to stick around for an ultrasound. I felt blood rush in my ears. This was probably nothing, I told myself. I was still so young, after all. The ghost of Trudy leaned over and whispered, "It might be something. Watch out." I twisted her ring on my finger and thought of my mother, just a few years older than me, sitting in a waiting room like this in Texas almost thirty years ago, tensed up, wondering, like me, if it is nothing, prepared for it to be something.

Someone called my name, and I lay on the table. We made a pretense at modesty, leaving my nether regions covered, but obscenely exposing my breast and smearing it with lubricant jelly. The technician watched the screen while firmly pushing the wand over me. "There it is," she said. She took photos, and then handed me a tissue to wipe off the lubricant. It felt strangely intimate and at the same time, perfunctory. "Stay here—we might need more pictures," she said, and then left to show this set to the doctor.

I lay there on the table, my sticky breast ensconced in the hateful pink robe, in agonizing contemplation. What if this was it? What if this was the moment when my life changed forever? What if the doctor walked back in here and had to tell me, at twenty-seven, that I had breast cancer? I pictured myself in a turban like my mother wore in that photo from the Southwest so long ago, only there was no baby in my arms. I had heard that chemo could induce permanent sterility. I pictured telling George and saddling him with a heavy burden so early in his career, my mother weeping, my life acquiring the quietness and resignation that comes when one is sick. I pictured dying with nothing to show for my life, dying without publishing a book or seeing Japan or eating, say, bull testicles. This probably wasn't that moment. But it could be, and what if it was?

It was a two-ham eternity later when the technician returned with the doctor, who told me they found a harmless calcium deposit inside me, and that they would follow their usual protocol and send off the films for a second opinion, as certain patterns of microcalcifications could be cause for concern. A preternatural calm came over me, and I managed to keep it together until I was in the parking garage. But when I got into the car I slumped over the steering wheel and let myself go to pieces for a few minutes. Then I drove to the gynecologist for ovarian screening.

Two weeks later, I received the official results in the mail. They arrived in a plain unmarked envelope that I mistook, at first, for a bill. There was a form enclosed with five checkboxes on it. We have examined your recent scan, it said, and the results are . . . The first check box, for benign results that indicated a completely clean scan, was empty, as was the last, for cancerous results that required immediate follow-up. One of the middle boxes, "probably benign," was checked. Probably benign. Probably. The thing in my breast was probably benign. Chances were really good that it was nothing. But probably benign isn't the same as benign. So, they did find something. Something that was probably nothing, but still something. There was something in my breast. I wondered if the hospital was trying to cover its ass for malpractice suits in case they got it wrong and it really was cancer.

These mammogram results infuriated me. Was this the future I could hope for, where a draining medical test yielded only nebulous answers? "Probably" didn't cut the mustard. I wanted certainty. I wanted assurance. I wanted to know that what was in my breast was not cancer and would not develop into cancer. I envisioned a future in which "probably benign" was all I could hope for, every six months like clockwork, bracketed by nervous anticipation. Horror rolled over me. I could keep my breasts and endure surveillance, but living with the crushing fear of becoming like my mother and grandmother, well, it didn't seem like much of a life to me.

The weight of breast cancer risk and her second biopsy in five years put my aunt Cris over the edge, just before that fateful Christmas so long ago. "I just think I couldn't go on with that fear and uncertainty," she tells me later. "And let's face it. By then all of my aunts on one side, my younger sister—I mean it just seemed inevitable, you know. It would be a miracle if I didn't get this cancer. You know, it wasn't like playing the lottery; it was for sure the right thing to do. Alan was just incredibly supportive. He never indicated anything less than one hundred percent support."

Soon after receiving my mammography results, I pressed George on what he thought I should do. He had been giving me extra hugs, shaking martinis, letting me pick what we ate and what we watched, and doling out neck rubs with extraordinary generosity, but he hadn't weighed in with advice. This was unusual for him—he is a man of strong preferences for everything from sock length to noise artists to bologna brands—and he is an especially strong proponent of science and science-based medicine. I assumed he had an opinion about what I should do based on data. At the same time, I feared what he might say about mastectomy, which had now edged out surveillance in my shaky affections. I was afraid he would tell me not to do it, because it meant cutting off a body part we both enjoyed. I had unanswered questions about my sexual attractiveness without my breasts, and about my sexual enjoyment. If he would miss them too much, how could I face him afterward? At the same time, I was scared he would tell me to go ahead with the operation too. I had this gut feeling that mastectomy was the right decision, but I really didn't want to lose my breasts. And I was secretly looking for an out, some overlooked justification that would permit me to feel OK about keeping them forever and ever. If he told me to go ahead, I would no longer be able to pretend that was possible. I was frightened to ask the question, and scared of his answer.

When we discussed it, I had to push him because he didn't want to weigh in. Years later, he'd tell me that he was afraid to influence me, afraid of that responsibility, afraid that his preferences in that moment could sway me from my own desires. He couldn't bring himself to say the word "mastectomy." "How do I feel?" he said when I pressed him. "I love you, and I don't want you to die." I pressed him further. "I would do the thing that has the best statistical chance of reducing breast cancer."

"So, mastectomy?" I said. It was still a question. It was not too late to avoid the surgery, but once he answered, I knew there would be no going back. He looked away from me and nodded.

"All right then," I said. "It's settled."

9 | Barbie Girls

A week after I decided to cut off my breasts, my grandma Meg died. She did not go gently into the night; rather, she went kicking and screaming, asserting her individuality up to the final moment. The last time I'd seen her, we'd come to visit, in part to introduce her to George. Though by that time she'd been having some mental problems, not just with memory but with anger, depression, and at times, viciousness, she'd turned on the charm for him, cracking jokes about the political situation and dazzling him with her smile. When we'd gotten engaged, she had sent us a red wooden salad bowl and tongs and made a little joke about how she knew we were living together, though we were not married, and she thought that was OK. She'd been a proponent of gay marriage because, she said, God is happy whenever two people who love each other decide to make a commitment. She thought people who leave their wives during breast cancer treatment should go to hell. "If she would have seen Newt Gingrich on the street," my grandpa once told me, "I think she would have spit in his face." She'd also been into fashion. A few weeks before the end, my mother remembers, they'd gone shopping together. Ever the fashion plate, my grandmother hadn't been able to decide between two cute outfits, so of course, my mother tells me, "I said we'd have to get them both." In the kitchen, her survivor hat, the cap she wore to Races for the Cure, still hangs. It is covered with pink ribbons, around forty of them, one for each year she lived after diagnosis.

I am relieved because congestive heart failure killed her, not cancer. Not cancer. Not fucking cancer. The cancer didn't kill her. She survived. It is such a relief that she survived.

At the memorial days later in her hometown of Fayetteville, Arkansas, I comforted my mother and managed to read a psalm aloud during the service. We were in full Rom family mode, with our emotions locked down while we were in public, though if you knew where to look, sorrow and relief etched everyone's faces. There would not be a home-cooked standing rib roast dinner for twenty-five served on fancy china, as there was after Gov's funeral a month earlier; the Roms are more pragmatic—they are tacticians with stiff upper lips—so we stopped at the store for cheese plates and potato salad, something easy, so we could be with each other instead of in the kitchen. Kathy and Lisa had flown in from the Midwest, tall and blond with serene faces, and since they had not been there for the wrenching end, they evened us out with pleasant memories of Meg from their childhood visits. My grandpa Roy has a serious photography habit and a secret sentimental streak. He covered the living room in framed portraits of my grandmother smiling mysteriously in front of the house, with the kids on vacation, or in some of the places they visited during their travels. You could tell what era it was by the hair and the clothes. I loved the smile on my grandma's face in one picture in particular. My grandpa told me it was from a visit to Italy they took in the 1970s. In it, she sits on a patio, leaning forward, with a glass of wine in her hand. Her greying hair floats in a cloud around her cheeks, and she's smiling and relaxed. Her flowery, flowing peasant top emphasizes her delicate build, and it seems like she is not wearing her prostheses, perhaps because it is the era of flat-chested mod fashions.

The day after the funeral, my aunt Cris and I ran a few errands together. In his grief, my grandpa wanted many of Meg's effects quickly sorted and donated, so we went to the women's health center at the University of Arkansas to donate my grandmother's prostheses in hopes some other, possibly low-income woman might get some use out of them. The prostheses, silicone things placed into pockets sewn into special bras, cost around $100 each.

The Halsted mastectomies my grandma underwent in 1968 and 1978 saved her life, but left her chest caved in. Without the prostheses, she looked as flat chested as a paper doll. Worse than flat chested, really—Halsteds made some women's chests concave. Luckily for my grandmother, the woman who put breasts on children's dolls would also put realistic falsies on women with mastectomies.

Barbie creator Ruth Handler developed breast cancer and underwent a mastectomy in the 1970s. As she wrote in her autobiography, the experience changed her self-image. "Losing a breast made me feel dewomanized. I'd been proud of the way I looked. I was well built and my designer clothes showed off my body. Now I felt the surgeon had taken the part of me that made me feminine and attractive. . . . I'd like to chop off parts of that doctor." Afterward, she wanted a little something to fill out her bra, so she started asking around. Her doctor told her to take some balled-up stockings and put them in that side of her bra. That didn't satisfy her need to look natural, so she went to a department store and asked. The salesperson took her to a dressing room and handed her a bra and a pair of gloves. Handler stood, confused, in the dressing room until finally she figured out that she was probably supposed to stuff the gloves into the bra. At a lingerie shop in Beverly Hills, she asked for a prosthesis, and the staff treated her request like a shameful, illicit deal, speaking only in whispers. The egg-shaped sack of liquid they handed her looked weird in her bra. In fact, the prostheses weren't even sized in conjunction with bras. Handler found this idiotic. "Every woman knows that her two breasts are as different as her two feet," she wrote. "We wouldn't think of putting the same shoe on both the right and left foot, would we? It was obviously designed by a man who didn't have to wear it."

So Handler decided to remedy the problem of bad prostheses and sales staff who didn't know how to treat customers with mastectomies. She had a prosthesis-maker cast her chest, and after a few tries she'd developed a more winning breast substitute made of foam and silicone. She founded Nearly Me, Inc., a company that sold prostheses sized to fit bras, that came in left and right versions, and weighed the same as the absent breasts. Ruth and a team of eight middle-aged breast cancer survivors marched into department stores where they trained salespeople in how to fit new survivors with kindness. Ruth even fit First Lady Betty Ford with a prosthesis after her mastectomy.

After the woman at the health center brought the receipt for my grandma's prostheses, Cris and I went back to the car.

"You know, Lizzie," my aunt said. "It occurs to me that this might not have been the easiest stop for you." It wasn't. I'd been carefully thinking light thoughts, reading flyers tacked to a bulletin board near the waiting room, looking forward to lunch later with all the relatives. For the first time, my aunt and I began to talk about the family curse. "How are you doing with all this?" she asked. "Do you feel crazy? Because I was a wreck in the months before my mastectomy, and it's very normal."

I did feel crazy. I felt crazy, desperate, and out of control. I often started and ended the day thinking about what had happened to my family and about what would happen to me, and I'd cry. Thinking about my breasts made me cry. Trying not to think about them also made me think about them, which made me cry. On breaks, I'd cruise cancer websites and read other women's stories, which were so sad I'd cry for them too. In fact, just the word "cancer," well, that could do it. It was stupid. It was stupid to be so sad about something that wasn't even cancer. Compared to my mother and grandmother, I was getting off lightly. I didn't deserve sympathy. I was a pathetic excuse for a human being. And that also made me cry.

My aunt's statement—that she'd had an emotional time before her surgery—felt like a revelation to me. During the ham Christmas, she'd probably been having a lot of feelings, which meant that I wasn't a unique irrational snowflake, a freak, a mutant drama queen. My aunt's acknowledgement of the fraught nature of this waiting period gave me permission to let up on myself. It felt freeing.

On our way home to my grandpa's apple orchard, we talked implants. I'd read that many women had only one regret about surgery—that they didn't go bigger. In contrast, I learned, my aunt went smaller and was very happy about that.

It's weird to think about selecting new breasts. If I go slightly larger or smaller, what does that say about me and my vanity? Will my new ones feel like breasts? Will I miss the old rack? The technical aspect yields more questions, because there are so many options. Do you go with implants? If so, silicone or saline? Or flap surgery transporting tissue from the back, stomach, or butt? Or a mixture of both flap surgery and implants? If you choose flap surgery, will the tissue get tunneled under your skin or cut free of your body and reattached with microsurgery? Will you keep your nipples, have them tattooed on, or rebuild them with darker skin from your groin, or will you stick with nippleless "Barbie-breasts"? Because this is kind of permanent.

Surgeons offer a vast number of procedures today, but it wasn't always so. The discipline of plastic surgery grew out of the ancient psychological need for wholeness, for love, for social acceptance, needs that found modern, narrower expression in the individualism and consumer culture of the 1960s and 1970s. And of course, as always, war drove many of the technical advancements that permit today's procedures by offering up plenty of victims with the desperate desire to regain a normal appearance by any means.

The surprisingly long history of reconstructive surgery begins with the Indian doctor Sushruta, sometimes known as the "father of surgery," who lived sometime between 600 and 800 BCE. An important figure in the history of medicine because his surgical texts traveled widely after being translated into Arabic, he made many advances—including the use of wine and marijuana as a primitive form of anesthesia. He also had a method for reconstructing noses out of patients' cheeks, cutting skin flaps that he sutured together to make new noses. As early as 1000 CE, Indian surgeons cut inverted spades of skin from their patients' foreheads and twisted them down to make new noses, leaving the points attached to preserve blood supply. The forehead flap, passed down through the generations, became known to English surgeons in 1794 and is still called an Indian flap today. The numerous street fights in Bologna, Italy, in the late 1500s damaged faces and inspired Gasparo Tagliacozzi, who was familiar with Sushruta's work, to create a new technique for nasal reconstruction. He made cuts around the nose and on the patient's upper arm, then stuck the patient's arm to his nose, immobilizing it with a steel contraption that looked like a torture device. After a few weeks, when the graft had taken, he'd cut the flap free from the arm and fashion a new nose.

Violent street fights propelled Tagliocozzi's new nasal advancement, and much later, the carnage of World War I would accelerate the field of plastic surgery. The war yielded an unprecedented number of facial injuries. Bombs dropped into trenches exploded upward into heads and necks. The steel helmets meant to save lives sometimes turned lethal, blown apart into fragments that hit comrades' faces along with projectiles. Early airplanes were dangerous and maimed pilots and passengers during crashes. Moreover, it often took days for the injured to arrive at hospitals, further complicating treatment.

The British Army had only fifteen dentists at the beginning of the war, so US ambassador Robert Bacon suggested to the president of Harvard that the university sponsor a medical unit. Columbia and Johns Hopkins joined in, sending thirty-five doctors and surgeons, three dentists, and seventy-five nurses over. The injuries they saw shocked and horrified them. Varaztad Kazanjian, a dentist with only two years of the experience, rose to the challenge. He wired crushed jaws together, improvised facial splints, and figured out how to put rubber inside people's faces so their wounds wouldn't contract until they got somewhere where bone grafting could be attempted. They called him the "miracle man of the Western front."

The war drove other advances in plastic surgery too, in the field of skin grafting. Ear, nose, and throat doctor Harold Delf Gillies, one of the fathers of modern plastic surgery, wrestled with a core problem of the discipline: it's easier to nip and tuck than sculpt new features. Gillies, working in Sidcup in 1917, and Vladimir Petrovich Filatov, working in Odessa in 1916, independently pioneered a new grafting technique, the tubed pedicle. Surgeons cut a long U-shaped portion of skin from the patient's body near the intended graft site and then stitched the long sides together to make a tube. They attached the end of the tube to wherever new tissue was needed for reconstruction. While the new skin took, patients looked like freakish aliens, with long, weird fleshy things dangling from their faces, but Gillies wrought astonishing transformations—turning monsters into men. Though now it's been replaced by vascular surgery and microsurgery, the tubed pedicle represented a major advance; it reduced the likelihood of infection in the pre-antibiotic world by keeping donor tissue enclosed by skin and helped maintain the blood supply to the new tissue. It also meant Gillies could rebuild noses and other facial features that had been blasted off, giving horrifically disfigured soldiers new leases on life.

During the war and after, Gillies trained scores of surgeons—mostly Americans who sought him out as a teacher. My countrymen dominated the emerging field, perhaps because it appealed to our vanity, or because in a classless society appearance spoke silent volumes about parentage, and in the great melting pot, it didn't do to look like the new immigrant. By the beginning of World War II, we'd have about sixty practicing plastic surgeons—ten times as many as Britain and almost double the combined total of the rest of the world, according to historian Elizabeth Haiken. By the early 1920s, we'd published the first respected plastic surgery textbook, and two American medical schools had departments dedicated to the new discipline.

After the war, when most of the blasted faces had been put together again, a glut of new specialists met a dearth of patients. The specialty would have to unearth a fresh market in order to survive. They had fixed disfigured men. Perhaps they could improve on the physically imperfect, a category that would, by the next century, come to include almost everyone. In 1921, these two threads—the new surgical discipline and an unattainable standard of beauty—began to intertwine. In August a group of doctors that would later become the American Association of Plastic Surgeons met in Chicago for the first time. A month later in Atlantic City, women vied for the title of the fairest at the first Miss America pageant. Yearnings for physical perfection no longer had to remain mere thoughts—as psychiatry entered the American consciousness, these longings became disease, inferiority complexes that the scalpel could slice through, offering a physical cure for a mental ailment, as surely as mastectomy attempts to cure fear of breast cancer in BRCA patients.

Though philosophers might opine that beauty is eternal, breast fashion says otherwise. Between the Middle Ages and the Renaissance, the ideal breasts were "small, white, round like apples, hard, firm, and wide apart," according to historian Marilyn Yalom. Breasts telegraphed social status. Upper-class women could afford wet nurses and thereby avoided stretching their breasts, while large bosoms afflicted the poor, who breast-fed their own kids and served as wet nurses for the wealthy. In the last century, the vogue has shifted every few decades—small breasts for the boyish 1920s fashions, torpedo-shaped large busts to welcome the men home from World War II, and then flat-chested again for the waifish fashions accompanying the feminist revolution of the late 1960s and early 1970s. After the push-up bra made inroads into the underwear market, large racks have enjoyed a resurgence. As a 1988 *Wall Street Journal* article proclaimed, "Breasts are back in style."

Surgery has followed the fashions. In the 1920s, cosmetic breast surgery focused on reductions and lifts for women with professions like dancing that required a certain body shape, and for women who couldn't enjoy normal stuff like sports without discomfort. By the 1950s, with the flat-chested flapper fashions long gone, the breast took on expanded meanings. US soldiers received "morale-boosting" pinup photos during World War II featuring women with prominent busts. The "breast fetish" and breast-centric fashions of the war and post-war periods, Yalom argues, "corresponded to very basic psychological desires" and carried clear messages for men and women. The breast-centric culture reassured men that the nightmare of war had ended and offered them the bosoms they'd dreamt of, while also keeping women in their place. They let women know, Yalom writes, that "your role is to provide the breast, not the bread. . . . It would take another generation to contest this vision of gendered fulfillment."

After World War II, the growing medical expertise of American plastic surgeons collided with rising acceptance of psychology and Freudian theory, leading patients to seek physical treatment for psychological conditions. The fad for big breasts made small ones into a disease; plastic surgeons said that women on the Itty Bitty Titty Committee suffered from "micromastia" or "hypomastia," wonky names suggesting an exotic tropical malady rather than a state within the normal variation of human bodies. At the same time that small breasts became an illness, the longing for a Marilyn Monroe rack became a mental syndrome. Psychiatrists described typical enhancement patients as "psychologically healthy-appearing individuals whose self-confident exteriors masked their true feelings of inadequacy, low self-esteem, depression, and neurosis," according to medical sociologist Nora Jacobson. Flat-chested women suffered from inferiority complexes.

Plastic surgeons and psychiatrists fed on one another. Surgeons recommended that women seeking augmentation undergo psychotherapy first, which meant more clients for shrinks, and also established among psychiatrists "the idea that surgery could be a legitimate treatment option for certain kinds of emotional maladjustment." Dramatic physical operations could cure mental malaise by revising the body itself. Or, to put it another way, it's easier to slice yourself open than to alter the prevailing beauty standard. The hell of it is that once you've altered yourself to fit a narrow definition of beauty, you become part of that oppressive establishment.

Technology cannot keep pace with women's yearnings. After my mastectomy, I didn't want something that merely looked like my missing breasts—I wanted exact replacements, down to skin sensation and breast-feeding ability. I didn't want to remove my breasts at all, just their ability to get cancer. But since that was an impossibility, I would settle for the best that plastic surgery could give me, an operation built on more than a century of mad science performed on the willing bodies of other women.

One of the earliest breast reconstructions ranked pretty low on the crazy scale. In 1895, Heidelberg surgeon Vincenz Czerny removed a tumor from a woman's breast and filled in the hollow with some fatty tissue from her hip or butt. A year later, her breast still looked good. Though today you'd sue the surgeon who gave you a uniboob, in 1903 Hippolyte Morestin described a reconstructive operation on a mastectomy patient that cut part of her remaining breast free and stretched it across the chest to make a single mound. Louis Ombrédanne described a technique for immediate reconstruction following a mastectomy in 1906—he flipped one of the minor chest muscles around, reattached it in a mound, and then covered it with skin. In the 1920s and 1930s, some surgeons continued Czerny's experimentation with fat not as a method of reconstruction but for augmentation, finding that the body absorbed it unevenly, leaving scars at the donor site and lumpy uneven breasts that looked ugly and made breast cancer diagnosis difficult. By the 1940s, transplants included skin as well as fat from the donor site to help ensure the graft didn't go necrotic, but the results were similarly unacceptable. Though a vast cavern of medical advancements separates these operations from today's, the basic idea is correct—it's possible to use a patient's own tissue in breast reconstruction.

But we didn't come here, back more than one hundred years, to hear about good ideas. We're here as voyeurs, to gawk at the two-headed fetal pigs in jars, the medical procedures relegated to the creepy basement with the gimp and that eugenics book inherited from our racist ancestors. Surgeons in the early to mid-1900s tried to put a lot of random stuff into women's tits, including glass balls, ivory, ground rubber, ox cartilage, and polyester wool. But for the real catalog of horrors, we've got to look at injectable fillers.

New York neurologist J. Leonard Corning and Viennese doctor Robert Gersuny kicked things off in the late 1800s with paraffin. Paraffin comes in two varieties, a stiff, waxy moldable version and a form with the consistency of petroleum jelly. Heat them, and they turn into a liquid. Gersuny diluted his melted soft paraffin with olive oil and began injecting small amounts into his patients. The body would absorb the oil, his theory went, leaving particles of paraffin that connective tissue would form around, filling in hollows. The connective tissue, rather than the filling that the paraffin might provide, was his aim. By the turn of the century, doctors injected paraffin into faces to fill in wrinkles and correct saddle noses, into the scrotum to create a testicle in one case, and into breasts.

This turned out to be a terrible idea. Paraffin could migrate around the body and wreak havoc. One woman ended up with pus and bits of paraffin leaking out of the broken-down skin around her injection site. Her breasts turned bluish and required amputation. Paraffin also caused localized tumors and could get into the lymphatic system, as Karl-Heinrich Krohn noted in his 1930 case reports of women with breast injections. Years afterward, some of the women reported joint problems.

Paraffin wasn't the only injectable disaster. Silicone, a new synthetic rubber, began to appear in medical devices in the 1940s and came in solid, gel, and liquid forms. Historians can't seem to agree whether this next tale is apocryphal or not. As the story goes, silicone-based industrial coolant kept disappearing from the docks in Japanese cities during WWII and reappearing in the hands of cosmeticians, who injected it into their clientele—prostitutes who wanted bigger breasts to lure their American GI clients. In 1946 American doctor Harvey D. Kagan claimed he'd used industrial liquid silicone—Dow Corning 200 fluid—to inflate women's breasts, but the technique didn't gain more widespread purchase until the

1960s. According to a report by the National Academy of Sciences, sometimes the mixture included oils, concentrated vitamin D, talc, paraffins, and snake venom. ("Snake venom: because injectable silicone isn't dangerous enough.") To top it off, the report also includes a list of other horrific injectable enhancements, including beeswax, shellac, glazier's putty, and epoxy resin. ("Epoxy resin: for when snake venom feels too safe.")

By the mid-1960s, the silicone injection craze hit America both on and off the books. Famed San Francisco topless dancer Carol Doda underwent twenty weeks of expansion, receiving about a pint of loose silicone in each breast. By 1964, with reports of complications emerging, the manufacturer of the industrial lubricant used for injections—Dow Corning—voluntarily listed it as a drug with the US Food and Drug Administration (FDA). A year later, the agency classed the substance as such, and Dow Corning began manufacturing medical-grade product. Liquid silicone wasn't available on the general market, but groups could apply to the FDA for permission to use it in controlled studies. However, these studies looked at low-volume silicone injections for correcting scars or smoothing out facial wrinkles, not high-volume breast injections.

Like so many beauty treatments over the years—lead powder to whiten and corrode the face, belladonna to give you sexy bedroom eyes and blind you—injectable silicone had serious side effects. At a minimum, it caused droopy bosoms later on. It also got into the lymphatic system and formed lumps in the breasts that made cancer detection difficult and could require amputation. If accidentally injected into the blood, it could cause rare but serious problems like gangrene, pneumonia, blindness, and death. A 1974 *New York Times* account of a divorcée who traveled to Tijuana for illicit injections vividly described the results. "Her breasts turned purplish, began ulcerating, and became horribly misshapened from the shifting and clotting of the silicone fluid." The FDA had classified silicone injections as a new drug in 1966, ruling that it wouldn't be approved for use pending further studies.

The siren call of insta-plumpness still lures patients, according to a 2012 *Daily Beast* article, now in the form of silicone butt injections. Originally popular among trans women, the seedy hotel-room procedure broadened its reach to women in the general population, bolstered by popular culture's recent craze for leg meat and big booties. The complications still include lumpiness, migrating silicone, respiratory failure, blood clots, and death.

Injectable breast enhancement failed the most basic test of a successful cosmetic procedure: it killed and maimed people. Luckily for women, World War II spawned the creation of many alternatives. Just as the First World War precipitated a search for synthetic chemicals, including mustard gas that later ended up in use as chemotherapy, the Second World War spurred the development of many man-made elements that had potential as breast implants. An ideal breast implant would be chemically inert and would survive in the body unchanged for long periods of time, unlike human fat, which tended to absorb unpredictably, or liquid paraffin, which migrated around. And finally, the new plastics joined the medical milieu whose name they shared. Although surgeons tried the war-produced materials of nylon and Plexiglas in breast surgery, they ended up focusing on synthetic sponges made of polyester, silicone, or Teflon.

Surgeons tested the new sponge implants in animals first. A 1951 study put implants into dogs and found no adverse effects after eighteen months. A study published a year later found that foam implants caused cancer in mice, possibly due to some chemical property of the implant material. Still, by the late 1950s and early 1960s, a variety of sponge implants joined the market for breast enhancement. Women had problems with drainage and infection after surgery, which propelled advances in surgical procedures, as when plastic surgeon W. John Pangman tried squirting antibiotic solution into the implant pocket before installing the sponges. Unfortunately, sponge implants had a more intractable problem. The body invaded the porous surface of the sponges with fibrous tissue that contracted and shrank the implants—sometimes by as much as 75 percent—and hardened the breasts. Imagine the bloody difficulty of carving out those failed implants that the body had colonized. Despite some efforts to improve the devices—covering the sponges with plastic and different grades of foam—medicine abandoned the technology.

Breasts have a tactile pleasure—not just in how they feel to the woman who wears them but in how they feel to another's touch—soft, pliable, and warm, so the next chapter in synthetic bosoms is fitting. Surgical resident Frank J. Gerow worked late at the lab one night in 1959. While he waited for a blood sample to fractionate, he killed time by playing with the hospital's new toys—plastic blood bags that had recently replaced the old bottles. They felt kind of like breasts. He pitched the idea to his supervisor, noted plastic surgeon Thomas D. Cronin, who called Dow Corning to see whether they could manufacture a silicone bag to hold saline. By this time, silicone was used in other implantable medical devices—shunts for the heads of hydrocephalic children, for example—with no ill effect. Although there hadn't been clinical trials on humans, doctors considered the material safe and inert. Cronin gave six dogs implants and followed two of them for eighteen months. By 1962 they had a prototype in hand and installed two implants into Timmie Jean Lindsey, whom doctors recruited for enhancement when she visited a charity hospital to have some ill-advised tattoos removed. Thirty years later, she still had the original implants.

Dow Corning launched the Silastic implant line in 1963, based on Gerow and Cronin's design, at first for augmentation and then, two years later, for reconstruction. The Silastic model featured a rubber envelope filled with silicone gel and a mesh patch on one side intended to fix the implant into place without sutures. The first implant had issues—a seam edge that patients could sometimes feel through their skin and that could harden, and the mesh patch, which could cause scarring. Over time, Dow Corning refined the design: the envelopes became thinner, the gel looser and more like breast tissue; the seams disappeared, and the mesh patch shrank and then vanished altogether. Silicone implants had another advantage—they could be manufactured en masse and sterilized before shipping, unlike sponge implants, which doctors carved to fit the patient and sterilized themselves. Alternatives filled with saline came on the market but had the unfortunate effect of deflating. Soon, implants of different shapes and providing varying levels of perkiness became available, presenting new choices to women who went under the knife.

Trouble still lurked on the horizon, though. In 1976, Congress passed the Medical Device Amendments to the Federal Food, Drug, and Cosmetic Act, which granted the FDA the power to regulate medical devices. After more than a decade of meetings, the agency let manufacturers know they'd have to do studies to prove that implants met the "safe and effective" agency criteria.

At the same time, as the new operation became more common, complications emerged, most commonly in the form of capsular contracture. The body recognized silicone implants as foreign material and sheathed them in scar tissue. This capsule could shrink over time, causing pain, hardness, and a weird look, as one surgeon told me, like a baseball inside a tube sock. Concerns about "gel bleed" emerged—the new thinner implant envelopes could permit silicone to seep through, a problematic concern given the legacy of silicone injections. Even more alarming, case reports linking implants to autoimmune diseases were published in the 1980s. Everyone geared up for a messy fight in 1991. Plastic surgeons and manufacturers defended their livelihoods, while women who wanted boob lifts and cancer patients seeking reconstruction straddled both sides of the issue. Surgery reinvigorated some patients who felt pleased with their new décolletage, while others had suffered complications, from botched surgery to capsular contracture—and a few had rheumatoid arthritis they blamed on the silicone devices. Medical societies suspected fewer women would seek treatment for breast cancer if the FDA banned enhancement, an assumption that sounds pretty sexist, alleging that feminine vanity would prevent women from seeking life-saving treatment. On the other hand, I suppose I can't judge since I might not have chosen mastectomy if I had known I would have had to live out life flat chested—the existence of reconstructive surgery made my decision easier. Was it a woman's right to choose her own procedure, or to choose procedures that had been proven safe? Why hadn't more studies been done?

Eventually, the government panel made an interim decision. Silicone implants would remain available to reconstruction patients participating in FDA-approved clinical trials. Augmentation patients would comprise a tiny portion of the studies and would have to stick with saline. No one liked the ruling because it distinguished between the needs of cancer patients and augmentation patients. If implant surgery fulfills a psychological need for bigger breasts, why should the desires of cancer patients trump those that nature made flat chested? Some cancer patients felt offended at the ruling's suggestion that, as historian Elizabeth Haiken put it, "their illness made them expendable, appropriate for potentially dangerous experimentation." If the problem was simply that no one had studied the long-term effects of implants, then the FDA's decision implied that women weren't smart enough to choose life over boobs unless the government stepped in and prevented feminine vanity from running amok.

Interestingly, the panel's decision left out testicle, calf, and pectoral implants.

Soon, the lawsuits exploded. In 1995 Dow Corning faced more than twenty thousand suits over its implants and more than four hundred thousand potential lawsuits worldwide. Fellow implant manufacturers Bristol Myers-Squibb, Baxter Healthcare Corporation, and the Minnesota Mining and Manufacturing (3M) Company also faced numerous suits. I was finishing middle school or entering high school when my mother received a letter notifying her of a class-action suit against the manufacturer of her implants and letting her know of her right to join the fray. Although she felt livid—enraged—that these devices had not been adequately tested, she declined to join the suit, since she hadn't personally experienced any adverse effects. In 1998, Dow settled what was then the largest class-action lawsuit in history, agreeing to a multibillion-dollar payout to tens of thousands of plaintiffs, even though no conclusive link between silicone implants and autoimmune diseases had been found. In fact, many years of science—including at least seventeen separate studies—has concluded that women with implants develop autoimmune diseases at the same rate as the normal population, although the FDA cautions that longer studies are still needed. Fourteen years after the original ban, in 2006, the FDA again approved silicone implants for reconstruction and augmentation.

The aim of all these strange materials, from wax to silicone to fatty tissue is, of course, to create a better breast. To the women who have them, breasts mean many different things—food for infants, markers of femininity, annoying things that get in the way of exercise, erotic sensation-givers, signifiers of adulthood, and aesthetic objects that look good in certain necklines. Plastic surgery concerns itself primarily with the last. The better breast, the ideal breast, is beautiful both clothed and unclothed. If beauty lies in the eye of the beholder, as the old saying goes, then the beholder is most often a male plastic surgeon, who has been historically inclined to condescend to his female client.

One 1976 plastic surgery textbook on breast reduction explained how they'd decided what "normal" breasts looked like. They took 150 healthy women and measured their breasts with their arms at their sides. "Of these, 20 were selected as being aesthetically perfect, or nearly so. They were regarded as normal," the book noted. Ah, the conflation of "perfect" with "normal"—so emblematic of American culture. If it's normal to be perfect, then most of us are freaks in need of plastic surgery. Also this construction of "aesthetically perfect" relies on the preferences of the people making the selections. If we'd gotten King Henry VIII to make the assessment, he'd probably have chosen the tiny-breasted. Even nowadays, both women and men have a variety of preferences—some view big breasts as ideal; others don't. It seems bizarre to consider less than 20 percent of the volunteer sample "normal." Since these ladies had similar breast dimensions that didn't vary according to height and weight, obviously "there exists a standard type of breast whose measurements would be aesthetically correct for any woman." One surgeon in the textbook went so far as to suggest that women shouldn't have a say in the size of their new racks because "they are far too emotionally involved." Oh, those crazy women, with their feelings and thoughts and desires about what sorts of surgery other people should perform on them. Decisions about breast size obviously require the superior judgment of men. Because who knows more about what it's like to have breasts than someone without them? Florence Williams, author of *Breasts: A Natural and Unnatural History*, reported that this literalization of the male gaze, etched into actual female bodies, stretched all the way back to the inception of silicone implants. Frank Gerow, half of the team behind the first Silastic implant, "reputedly liked big breasts, and apparently it wasn't unusual for him to take a look at an unconscious woman on the operating table in whom he had just placed implants, decide she could handle bigger ones, and redo the whole thing."

I'm not the sort of woman who dreamt of plastic surgery. I find the whole notion rather problematic, of expanding one's body to fit the prevailing trends rather than fighting for an enlargement of what counts as normal and beautiful. As a feminist, I find the idea that I must pluck, tuck, resurface, and slenderize my body abhorrent. Yet, I do many of these things because I have internalized our culture's body shame to such an extent that I feel I must do them to make my body acceptable to myself as well as others. Such beauty treatments construct a veil, a mask between me and the world, a false front that bricks my true self off, hides it in a secret compartment accessible only to me and a few permitted others. Yet the longer I wear the mask, the more it becomes me. Pay no attention to the woman behind the curtained black bangs. My constructed façade—cartoonish haircut, black-and-red uniform, fake breasts—is passably attractive. The real me, the secretly ugly me underneath the expensive makeup and concealing drapery—no longer feels like the authentic self it once did.

I wasn't always like this—before I cut off my breasts, I had some fight against the prevailing beauty standard left in me. But after submitting to the surgeon's gaze, after I received the eternally perfect bosom I never wanted, suddenly, it all seemed within my grasp—the ageless face, the hourglass figure, a flawless complexion achieved with neurotoxins, liposuction, and chemical abrasion. Before my surgery, I neither plucked my eyebrows nor shaved my legs regularly. I scoffed at plastic alterations—I am me, I am my body, and I am as good as anyone else just as I am. After my surgery, I fantasized about quick fixes to all the things I hated—my saddle bags, my spare tire and my angel wings, the incipient crow's feet, the jowly cheeks and loose under-neck beginning to emerge, even at thirty-one. I might have been young, once, before this ordeal, but now I feel old, trapped forever in a body with a youthful chest, the rest of me doomed to wilt and fade. I'm Meg's granddaughter, insecure about my appearance, worried that everyone who looks at me thinks about my mastectomy or thinks me vain for the surgery I never sought, anxious about the lie my faked cleavage presents to the public. I had a body, once, a natural body that existed in the world, but plastic surgery changed it. And in changing my body it changed me.

10 | Captain Kirk and Doctor Spock

I lost control of my body. In the six weeks after I learned of my BRCA mutation, I gained fifteen pounds—a shade over 10 percent of my body weight, and it was not on the wane. I was an undisciplined emotional eater, feasting on fatty cuts of stewed beef, baked goods smeared with cheese, ice cream, and cocktails. The reason I gained the weight wasn't a mystery, but my inability to pause my ravenous appetite, to eat carrots instead of pasta, to even attempt exercise, well, I found it curious. I had no spare energy with which to exercise self-control; all my energies had funneled into grief. I outgrew my pants and had to buy bigger ones. I ballooned out of those, and pride wouldn't let me size up again—I let the arbitrary number on the tag make me feel inadequate. It was part of how I punished myself for my feelings. I decided pants were a privilege of the thin, a class I no longer belonged to. Lately, as a function of the deaths in my family and my dark mood, I had been feeling the black clothing anyway, and I deceived myself into thinking that if I wore dresses, that if I wore black—it's slimming!—no one would notice.

But of course, people noticed. A Latina cashier at the supermarket stretched her hand toward my stomach and asked when I was due. A white lady in line at the CVS, with three kids tugging at her, asked me too. While I was at a convention in a hotel, the woman restocking the paper towels in the ladies' room struck up a conversation with me about my dress, and prefaced the question with, "Don't be mad, but . . ." It was most awkward in social situations—people I was slightly related to or people I'd just met at parties—and it was worse when someone reverently touched just below my breasts, which had gotten huge from all the weight gain. How do expectant women bear this invasion of personal space? I imagined using a machete to sever those hands at the wrist and then stringing them into a necklace I'd wear out in public. No one would ask me if I was pregnant then. These people meant well, of course—many of the unpleasant reminders came from immigrant women hailing from a different social context, who asked me these questions as if partaking in my supposed joy. Paranoia set in. Did that guy on the subway give me his seat because I was a woman and looked exhausted or because he thought I was pregnant? While I had been fielding the occasional question about what might or might not be happening inside my uterus, even when I was twenty-three and twenty pounds lighter, I loathed the constant reminders that my body was as out of control as my emotions.

The pregnancy questions hurt because soon I would give up breast-feeding. And while in the scheme of things it's a minor sacrifice, I felt angered that I had to give up anything at all, that the world's tiniest scalpels could come in, slice off a bit of my DNA, and tell me the future, while the solution remained the same draconian one that Frances Burney had endured more than two hundred years before. I also wasted energy feeling angry at myself, for my undisciplined eating, for my choice of food, for my fatness, and because, as a feminist aware of the discourse around women's bodies, I should have inoculated myself against this sort of self-hate. Congratulations, prevailing beauty standard, you won a pyrrhic victory. Intellectually, I knew that gaining weight wasn't immoral and didn't make me a lesser person, that I shouldn't let the extolled concept of skinniness get inside my head and make me feel inadequate. But we can't always feel as we should. When I looked in the mirror I saw an ugly lump, transparently trying to deceive the eye of the beholder with a black dress.

I found it difficult to be around people other than George. I did not have many friends where we lived in New Jersey—we had moved there because George was working on a PhD in biophysics. Most of his friends were scientists. I telecommuted into my news job but otherwise worked for myself, alone in a coffee shop. So when we went out, it was with his crew. I felt I couldn't connect with them. They had many fine qualities—smarts, intellectual curiosity, kindness, humor—but they made physicist jokes about angular momentum, and my humanities-oriented mind felt out of step, out of place. When I saw them at parties and gatherings, we made charming small talk, but any social interaction at all felt like fingernails on a blackboard. Emotion had rubbed me raw, and it exhausted and pained me to have to smile in public. We turned down a lot of invitations, and I felt guilty because I was poisoning George's social life with my misery. If we did not accept some, soon they would stop coming. Sometimes I sent George without me, but that had its price too—I went whole days without really talking to anyone, and when I was alone in the house at night I read up on cancer and wallowed in misery.

Eventually, I recognized my symptoms—the constant crying, weight gain, lack of interest in socializing, and reliance on a nightly cocktail as a sleep aid—for what they were: depression. A trip to the doctor netted me some antidepressants, and even though they took a few weeks to really kick in, the simple act of having sought help made me feel immediately better. Soon, I was only crying once a day, and then only a few times a week. It was a relief to not have to feel so miserable all the time, but it came at the price of distance. I felt as if I were observing myself through a pane of glass. I could remote control my own actions—but I was unmoored from myself, mind divorced from body, emotions from passions, and it felt unsettling. If I could not feel grief as keenly, then I could not feel joy as immediately either.

As I reached this place of tentative stability, I began searching for surgeons. The first stop was my oncologist, who let me know that, generally, teams do these operations. I would need a surgical oncologist, who would cut into my breast and remove the tissue, as well as a plastic surgeon, who would reconstruct my breasts. My oncologist gave me advice I'd later heed, namely that he had found that, of the two, it was more important for women to click with their plastic surgeons, that this personal relationship, with the person who would, after all, be creating the aesthetics of your new body, was more important than your relationship with the person destroying your old one.

I knew from my research on the Internet that techniques for mastectomy and reconstruction had advanced quite a lot, even since my mother's time. For starters, the surgical oncologist could use many different sorts of incisions to remove my breasts but leave the skin intact. There are oblong incisions shaped like an eye, where the nipple would mark the pupil. There are incisions named for tennis rackets, anchors, and keyholes, and of course, it is also possible to remove the breast through a simple small incision in the crease where the breast meets the ribcage, a place called the inframammary fold. Once the cut has been made, it's possible to remove the breast tissue with a scalpel, as if skinning you from the inside out, or with a cautery iron. Finally, the surgical oncologist may or may not remove the nipples.

That's the first thing I decided—that if it was possible, I would like to keep my nipples. Keeping your nipples isn't standard, because nipples aren't skin—they're breast tissue, the place where milk ducts exit the body. Since the point of this surgery was to remove as much breast tissue as possible—though no surgeon could get every single breast cell—it made sense to remove nipples along with everything else. But I stubbornly wanted to keep my nipples, in part because I thought it would help me feel more "me" after the surgery, but more importantly, because some women who did nipple-sparing surgery maintained some skin sensation in them. Removing the nipples naturally guaranteed loss of nipple sensation in addition to the numb skin mastectomy creates in certain parts of the breast, since the operation cuts through nerves as well as flesh. Since nipple sensation was important to me, the risk of keeping a small bit of breast tissue in exchange for the benefit of possible nipple sensation was one I felt willing to take. There wasn't really science on whether this was a good idea or not, but my oncologist passed me some papers suggesting that breast cancer almost never starts in the nipple, and that was enough to make me feel vindicated, justified, in choosing to spare mine.

I decided that I'd also like my incisions to be in the inframammary fold and nowhere else. That way, my breast fold would hide the scars, even while I was naked.

Plastic surgery made the choices even more complicated. The two basic methods for breast reconstruction are implants and tissue flaps. Implant reconstruction uses synthetic materials—sacs of silicone or saline—to create a new breast, while flap surgery uses tissue from a donor site on your own body, like your butt. Each method has dozens of variations, benefits, and drawbacks. Implant surgery tends to be less complicated and have a shorter healing time but of course involves having foreign objects inside you semipermanently, which freaks some people out. And implants frequently need replacement five or ten years down the road, so there is a risk of repeated surgery. Flap surgery is more permanent, but it's also more complicated because it involves surgery to the donor site as well as to the breasts. Because of this, it has a longer recovery time and the complications tend to come immediately after surgery, when tissue can go necrotic and die.

Initially, I thought flap surgery sounded cool. Doctors could tunnel tissue from the donor site up to where it was needed, or they could actually detach it from your hips, ass, or stomach, and reattach it to your breasts in a technically complex but aesthetically satisfying procedure. If you gained or lost weight, your breasts would still fluctuate with that after some sorts of flap surgery. I liked the idea of taking the unwanted spare tire around my middle and turning it into breasts. A tummy tuck and a boob job at the same time—if the gene had depressed me into weight gain, maybe the magic of science could fix that at least. It sounded like the silver lining of this whole shitty cloud.

My aunt Cris had chosen flap surgery—to preserve the blood supply, her doctor tunneled stomach under her skin up to her breasts. She'd always been thin—in high school her parents had thrown her a party when she broke one hundred pounds—and her surgeon said she was the tightest person he'd ever worked on, in the margins of what would work for this procedure. The resulting breasts were smaller than her originals, which suited her fine. "I had always been a little top heavy," she says, "with scrawny arms and scrawny legs," and her new rack suited her better. After the surgery, the skin across her stomach was very taut and flat, thanks in part to the Gore-Tex screen embedded in there to strengthen it.

I booked an appointment with a local plastic surgeon, one of the few in the state who performed the free-flap procedure, called a DIEP flap, in which the surgeon detaches donor tissue from your stomach or hips, puts it up on your chest, and then reconnects the blood supply using microsurgery. Tall, thin, and dark-haired, he gave me the distinct impression of a used car salesman on the prowl for new customers. "Put away your notebook," he told me. "I usually find that patients retain information better the first time if I just give it verbally and they listen rather than writing it down." Oh really, Doctor? I wondered in my head. I didn't realize you were an expert on my personal cognitive processes. I am a reporter after all. Processing information while writing notes is how I make my living.

But I'm an obedient patient, so I put my notebook away and let the facts fly over me. He talked rapidly, explaining that all women really wanted was cleavage that looked reasonable in low-cut shirts, and he could give that to me—and to any friends I might have who were interested in boob lifts. We could get boob lifts together! he told me. When he heard my mother hadn't had revision surgery on her implants in several decades, he told me that it wasn't normal for a woman in her late fifties to have such perky breasts, and that if we came in together he could revise them so they drooped, just like they were supposed to.

A nurse came in to watch for malpractice suit protection—standard procedure—while he measured my naked chest with a measuring tape and squished my copious stomach fat down in his hands to get a sense of whether I had enough to make new breasts. They'd have to be smaller, he said, plus if I was planning on kids, the procedure might make my stomach too tight for pregnancy. I wasn't sure I wanted kids, but I definitely wanted to keep the possibility open. And it'd leave me with a scar from hip to hip that would be visible in a bikini. Normally he did free flaps on post-pregnancy women who had a pad of fat leftover from child-bearing that hung down in a fold where he could later hide the scar. No, for me, he recommended implants. Since I had very round breasts, he said, and traditional implants created a round breast shape, I was a good candidate for the procedure.

First, a surgical oncologist would cut out the breast tissue, he explained. Then they'd lift up my pectoral muscles and stuff something called an expander back there. It would have a port in it right where my nipple was. Keeping the nipples was not a great idea, he said, because of course they were breast tissue. Then over months, he'd take, as he put it, "my special water gun," hook it into my nipple port, and fill the expander gradually, so that my pectoral muscle would stretch out over time. Once I reached the desired size, he'd do a quick, in-office tire-exchange operation where he'd swap out the expanders for permanent silicone implants. Nipple reconstruction would come later. When could he schedule me? He had an open appointment in two weeks, and he could just call over to the hospital right now if I wanted . . .

His callousness and condescension upset me. I couldn't believe he'd demanded that I put away my notebook, tried to convince me that my friends needed breast augmentation, and said that thing about wanting to revise my mom's breasts so that they looked appropriately droopy. My mother is beautiful. Then he'd pressured me to have the surgery in two weeks. I could have the surgery in two weeks. The idea that it could happen so soon freaked me out. Indignation, irritation, and dread of the surgery combined within me, and when I got back to my car in the parking lot, I leaned over the steering wheel and had a good cry. My new custom. I decided that maybe I shouldn't be going to these appointments alone, but I hadn't wanted to pull George or a friend out of work to come. I'd been going solo, as I had to those tension-filled mammograms, because I didn't want these appointments to feel like a Big Deal. I wanted to pretend, to myself at least, that this was all completely routine and that I was a grown-ass woman taking care of her health and not flinching.

That afternoon, I called my mother to report on the meeting—she'd helped me come up with a host of questions for the doctors, and we had a lot of fun making sarcastic comments about this dude-bro. "Oh yes," she said. "I desperately want breasts that droop. How did he know?"

I certainly didn't want that dude reconstructing my breasts—he'd probably give me Real Housewife tits—but even if I did, as with all surgery, it would be wise to get a second opinion. I knew that some of the most renowned surgeons who performed the free-flap procedures and mastectomies on BRCA women operated in New Orleans, but frankly I didn't want to travel so far for surgery. The pregnancy complication thing and the long recovery time scared me. The DIEP flap has a six-week recovery time from the basic surgery because, in addition to having breast surgery, you're having abdominal or butt surgery. Nipple reconstruction and "revision" surgery—operations that tweak the final product—mean that it can take almost a year before your new breasts are complete. Most important, though, I wanted my new breasts to resemble my old breasts as closely as possible, and if there was not enough stomach fat, then there was not enough stomach fat.

I researched more doctors. I lived in New Jersey, not so far from New York City, where the famous Memorial Sloan Kettering Cancer Center is. I called them up. "I'd like to see a surgeon," I said, and listed some names. "I have tested positive for a BRCA mutation and want to remove my breasts." "You want to see a surgeon?" the woman on the other end said, as if I had asked to borrow a walrus on roller skates from the hospital. "Yes," I said. I explained that I wanted to have my surgery in about six months. "Why do you want to talk to the surgeon now?" she asked, her tone of voice suggesting that she simply could not believe my nerve. But still, Sloan Kettering is a good hospital, so I soldiered on. "Because I'd like to have a sense of the person operating on me," I said. "We don't do that," she said, "until right before the surgery." The irrationality she seemed to be projecting on to me made me feel alternately enraged and ashamed that I was causing trouble. Of course I wanted to meet the person who was going to hack off parts of my body well in advance of the operation, because although my risk was high, unlike an actual cancer patient, I had the luxury of time. She continued treating me as if I were making a crazy request. After another five minutes on the phone I discovered that they wouldn't let me see a surgeon at all until I had been to see their genetic counselors. I told her I had already done genetic counseling, that I had my test results, that I was ready for surgery. Nothing doing. The next open genetic counseling appointment was in four months. It felt too long, but I took it just in case. When I got off the phone, I started crying again. I didn't like being treated like a crazy hypochondriac. She was probably having an off day and doing her best. I was doing my best to hold on to the shreds of my sanity too, but even with the antidepressants, it was difficult.

By this time, I was addicted to the FORCE website. FORCE—Facing Our Risk of Cancer Empowered—is an advocacy group for BRCA mutants like me. The best thing on the website was the forums, where previvors—survivors of a genetic predisposition to cancer—could talk and give each other advice. I found the word "previvor" sort of creepy because I thought it implied that this gene now constituted a central part of my identity—and though it has shaped me, it does not define me. At the same time, it felt comforting to have a label, to know that although I felt like a one-eyed tap-dancing lizard, I was not alone in feeling depressed, angry, desperate over something that was, after all, really nothing. Though I no longer felt like one, I was still a healthy twenty-seven-year-old newlywed. I posted anonymously on the site and learned about the sorts of surgery women in my shoes have, and this was where I found out about the one-step, or direct-to-implant, surgery. Thought it's often billed as a brand-new reconstructive technique, the one-step with nipple sparing has been around in some form for at least thirty years. My mother's cousin Kathy had one after El's death in 1979. Essentially, the surgeons perform the mastectomy and complete reconstruction in one four- to six-hour-long procedure. The pioneers of a new and improved version of this operation, a two-person surgical oncologist–plastic surgeon team operated about two hours away from me and specialized in BRCA women. I called and booked appointments with both offices.

These appointments had become increasingly intense for me, and I didn't want to go it alone anymore. This was a Big Deal. I was getting ready to cut off my healthy breasts. So I called my mother and asked her if she would drive up from Washington, DC, to New Jersey and visit these doctors with me. After a mastectomy for cancer in one breast, a voluntary mastectomy in the other, a later bout of reconstruction, a thyroidectomy, and a hysterectomy, to say nothing of the gum grafts, my mother is pretty smart about surgeons and surgery. She's shopped around for surgeons before, and even fired a few. For my mother, bedside manner is important. The day after her breast cancer diagnosis, she remembers talking to the oncologist on call in the hospital, who answered her first question, "Can I have children?," with a blunt "No," and her second question—"Is the whole tumor cancerous?" (since she, still shocked from the diagnosis, thought perhaps it could be a cyst with cancer in the middle)—with a sharp "Yes." "In retrospect," she says, "he had to deliver bad news to me," which probably wasn't pleasant for him either. Still, the following day a new oncologist was on call. He came into her room and struck up a conversation about the pretty flowers someone had sent. My mother loves flowers. That little human touch, the kindness to her psyche, made her ask to be switched to become his patient.

My mom's search for a plastic surgeon followed a similar trajectory to my own search. The first guy she went to "said that he wouldn't do the reconstructive surgery on me," she says. "He said that all my energy was going into fighting the cancer and that, if he did reconstruction on me, I would just get cancer again." When she left the hospital she was crying so hard that the parking attendant asked her whether she was sure she could drive. As my mother learned when she brought up the conversation with her oncologist, that plastic surgeon was wrong about reconstruction driving cancer recurrence. Her doctor let her know about a top reconstructive surgeon in Atlanta named John Bostwick III. This was in 1983, long before Internet searches made research so easy, so my mother travelled to the local medical school and looked up this Bostwick character in the library, read some of his papers, and read medical texts about the procedure. Then she traveled to Atlanta for the day and interviewed him. He worked, as many surgeons then did and still do, in several stages. After the implant process, three months later she'd have to return for nipple reconstruction with darker skin from the groin area. "Lordy, I never thought I'd be writing about this with such casualness," my mother e-mailed me just before our visit. My mother's reconstructive surgery was more complex than mine would be because she had radiation damage from the cancer treatment. On the side where she had a prophylactic mastectomy, her implant is placed below her pectoral muscle. On the other side, Bostwick tunneled one of her back muscles under her skin to make a pocket to hold it.

The decision to have reconstructive surgery had come easily. "I had worn the prosthetics for the year," she says. "And they're hot, and they feel heavy; they don't feel like a part of you. I was going to the gym and stuff like that, and I didn't want to wear them to the gym; and I thought, well, why should I have to do without this if it can work for me?" The reconstructive surgery also helped her recover after the mastectomy, which had been "a sexual loss.

There's no doubt about it. As I recovered from the surgery and the treatment from cancer, I was also able to recover a sense of sexuality and femininity, and I was able to feel beautiful. That was one of the pleasures of having the reconstruction," she says. "I felt . . . not whole . . . and not 'beautiful'—that wasn't exactly it either," rather, she felt like herself again. "And I had that shape back that I was used to, and it felt like a part of me, part of the total me, not the physical me."

One sadness for my mother during her cancer was that she wasn't able to connect with her mother over their shared experience, in part because she opted to deal with the illness privately, with my father, and in part because my grandmother sometimes got wrapped up in her own experience. She told my mother that she hadn't needed reconstructive surgery to feel good about herself and didn't encourage the operation, which hurt my mother deeply. It was only decades later that my mom would come to understand that reconstruction hadn't been an option for Meg due to the extreme nature of the mastectomies and radiation she had endured. Meg's disapproval may have stemmed from her own disappointment. Perhaps she found it easier to deal with the reality of her unreconstructable chest if she convinced herself she wouldn't have wanted reconstruction anyway.

This was why I wanted my mother with me at these appointments: even before we met the surgeons, based on accounts I had read on the FORCE boards, I thought they might be the guys. With her experience, she would be an amazing second pair of ears and would ask good questions.

Aside from that, she is my mother. She would walk into a burning building for me. She would take a genetic test for me. And if it were possible, she'd have this mastectomy for me. Though we talked often on the phone and via e-mail, I could tell that she had been holding back, trying to give me space to deal with my own emotions and make my own decision, trying not to crowd me with her baggage.

But I wanted my mother here. It would be good for both of us, I thought, for me to be able to feel her love up close and for her to feel permitted to love me during this tough time. It can be difficult to let other people help you, to let the people who love you see you weak and in need of support.

When we arrived at the office of surgical oncologist Dr. Andrew Ashikari after our two-hour drive, we discovered that I wasn't in the appointment book there, or over at the office of the plastic surgeon, Dr. C. Andrew Salzberg, either. Perhaps there had been a mix-up with these two doctors, or perhaps, clouded by emotion, I had only imagined that I'd made an appointment on that day. In a move that would further endear these doctors to us, the receptionist managed to find a slot for us later in the afternoon and even called ahead to Salzberg, who was able to see us about forty minutes later.

Salzberg's office was in a clean modern building that resembled an obsidian box. The interior was tidy and new and painted in neutral golden colors. It felt like we were not in an office so much as in a day spa. The receptionist escorted us to a small room containing a large wooden desk with two armchairs across from it and offered us tea.

Dr. Salzberg, when he entered, reminded me of Captain James T. Kirk of the USS *Enterprise*. With his light brown hair, blue eyes, and sloping shoulders, he resembled a middle-aged William Shatner and had an easy, comfortable intimacy with us even though we had just met. He sat behind the desk as if he were in no rush and explained direct-to-implant surgery to us while we sipped our tea. We already knew most of the details because the receptionist had put on a short film for us, projected onto the wall behind the desk, while we waited for him. Essentially, after the breasts had been removed through an incision in the inframammary fold, the surgeon lifted up your pectoral muscles and inserted implants beneath them. The muscle covered the top part of the implant, and to hold the lower part of the implant and provide coverage, Salzberg inserted a sling made of AlloDerm, which was a synthetic collagen matrix. What in tarnation is that? I asked. It was essentially donated tissue, harvested from corpses, that had been stripped of DNA and sterilized. As you healed, your own tissue would grow into it, and a few years down the road, if you were to slice off a bit and DNA test it, it'd have your own DNA. Salzberg pioneered the use of AlloDerm to support the underside of the implant—an advancement since the time of Kathy's similar surgery. This procedure had the advantage of completing reconstruction and mastectomy all in one go. The immediate recovery from the operation took only a few weeks.

Dr. Salzberg joked that he saw Dr. Ashikari, the surgical oncologist, more often than he saw his wife. The two of them specialized in nipple-sparing surgery for BRCA women and had done hundreds of operations. We paged through a binder of before and after shots of headless naked women. Some had had mastectomies of one or both breasts without reconstruction before they came to him. They were fat and thin, old and young, but afterward their breasts all looked the same: bouncy and preternaturally round, taut and shiny. He pointed out the surgeries he was happy with, and the ones he wasn't. We looked at shots of capsular contracture, the most common complication of implant surgery. Over time, the capsule of scar tissue the body surrounds foreign objects with can shrink, which makes stricken breasts look weirdly bulbous and taut. Salzberg explained that he believes that AlloDerm helps with capsular contracture because it has a little more give to it than the pectoral muscles that cover the implant through traditional reconstruction with expanders. Of course, he said, he could do traditional expander reconstruction on me if I preferred, though he usually preferred expanders for women with more complex health issues, for example, tissue damage from radiation for cancer. He took some implants out of a box and handed them to my mother and me to touch. I was fascinated by the viscously smooth texture. As I rubbed two sides of the envelope against each other, it felt like some sort of expensive office toy. We were ginger about it too, though, playing with them only for a moment before handing them back, as if it wouldn't do to seem too interested.

In the exam room, Salzberg took some measurements of my chest while a nurse watched, and pinched my stomach fat at my behest to confirm that no, I probably did not have enough for a DIEP flap, unless I was willing to go smaller. I was not. I wanted as little to change about my body as possible, though I had interrogated some women on the FORCE boards, and it seemed like the main regret about this procedure was not going a bit bigger. I would vacillate about this a bit. I had never wished for a larger endowment than nature gave me, but if most women ended up disappointed, well, perhaps I might too. I had no idea how I would respond to having a new body. Salzberg handed me off to an assistant who had me stand topless on a line on the floor while a large 3-D camera mounted on a metal T moved robotically up and down, taking an image of my chest. On the computer she showed me what I would look like after the surgery from various angles, digitally trying out different sizes and shapes of implant. I've been in your shoes, she told me. She was thin and tall, young and pretty, with dark hair and a slim-fitting purple dress. From looking, you'd never know that she was also a mutant like me, and that she had had the very same procedure I was contemplating. She looked . . . perfectly normal. She was not the only former client who worked for Salzberg. The beautiful post-surgical doctor, who worked with patients to help optimize the cosmetic outcome afterward, had also had this same surgery. In my mind, these were powerful endorsements. I liked Salzberg's no-pressure attitude, his willingness to show us his mistakes, and his bedside manner. I was surprised how much the latter meant to me—surgeons see lots of patients, and it can become routine, I suspect. But for each patient the experience of surgery can be traumatic and upsetting—I liked that he had not slipped into routine.

Though it had been informative, this visit had not been easy. We would be seeing Ashikari, the surgical oncologist, a few hours later. My mother and I decided on Mexican food, followed up by ice cream in the quaint local town. We both liked Salzberg—his William Shatner–ness and the confidence he projected, how he had explained the procedure to us and answered all of our questions without condescending. Perhaps in comparing him to Shatner, I'm retelling the old story about plastic surgery, making him into the hero and me into the silent, passed-out damsel. But it felt like such a relief to find someone who seemed both kind and competent. And more than anything, we liked the tea his assistant had served. It was just the thing, we said over and over again, using different words, to calm down two people in the throes of surgical emotion.

Ashikari's office didn't feel as posh as Salzberg's, but it was busier, in part because Ashikari treated cancer patients as well as otherwise healthy BRCA patients, and there's an urgency to cancer treatment that there's not to, say, using ultrasound to tighten and tone loose facial skin. The office, where he practices with his father, who is a renowned oncologist, has nearly ten thousand active patients, who need care and follow-up during the long road of cancer treatments. In an unassuming spot on one wall, I read a framed poem, apparently written by one of the Ashikaris' patients, in which she prays to God while suffering from cancer, and then looks up from her bed to see his presence in Ashikari smiling down at her. It wasn't Audre Lorde—it rhymed "Ashikari" with "see"—and was neatly handwritten, presumably by the writer, a clearly heartfelt testament to the tremendous faith one patient had in the medical care provided by this office.

Dr. Ashikari was tall and thin, with a long, narrow face and dark hair—the perfect Spock to Salzberg's Kirk. And as it turned out, though he was a nerd's nerd when it came to the science of surgery, he also had the cordial social graces of a debutante. He saw us in his office, behind a large desk that was covered with stacks of medical papers he was clearly in the process of reading. Although his receptionist had shoehorned us in, he treated us as if we were his only patients and gave us thorough, unrushed answers to every question. He talked about the papers on nipple sparing that he and Salzberg were coauthoring.

By 2013, a few years after I saw them, he and Salzberg had performed 417 nipple-sparing prophylactic mastectomies, more than any other single practice in the country, making Ashikari, according to his bio on the Ashikari Breast Center website, the world's foremost expert. About two-thirds of those 417 women were positive for a BRCA mutation, while the rest had made decisions based on other risk factors; some were cancer patients who only had a prophylactic mastectomy in one breast. As of the beginning of 2014, Ashikari told me, only one of those patients had developed cancer, a woman with a strong family history but not a BRCA mutation. After the mastectomy, the pathologist had found atypical tissue spread throughout both breasts. We talked about AlloDerm and how Ashikari and Salzberg had tried out operations in a few patients using a collagen matrix made from pigskin but found that in one of them, the pigskin liquefied, requiring the removal of the implants. I sat, feeling queasy about liquefied pigskin inside some poor woman's breasts and happy that I would be using something else. He referenced the latest studies and periodically looked them up on his computer when he was in doubt. Clearly, he viewed his profession as a living thing that he watered with frequent study.

He talked about how sometimes women came to him wanting, essentially, a free boob job from their insurance, and how he treated these women with skepticism because it wasn't a boob job. It was also a mastectomy. He thought I was the sort of person who had accepted the reality of the situation, and he was not worried about me, he said. I felt proud that I seemed like a person who was dealing with this rather than falling apart, but I also dreaded being ready for the surgery.

Although this surgery had a cosmetic angle—I didn't think I'd be able to cut off my breasts if I weren't getting new ones—really, the mastectomy was the crucial piece for cancer reduction. Dr. Ashikari explained that inside the operating room, his main goal was keeping me healthy by removing as much breast tissue as possible. Dr. Salzberg would sit in on the whole surgery, watching Ashikari make the initial incisions, and commenting. If Ashikari could help the cosmetic outcome by making the initial cuts a little to the left, then he would. But, he said, he would not compromise my health. If Salzberg said, "Can you make the cuts a little smaller?" and Ashikari needed them bigger in order to remove all of the breast tissue, then he would make them a little bigger. If Salzberg wanted him to take a little less tissue to make the skin flap thicker to make the cosmetic result better, Ashikari wouldn't do it. Inside the OR, he would be the cancer doctor, removing as much breast tissue as he could. As he told me, "My judgment for a prophylactic mastectomy versus a mastectomy for cancer is really the same. I'm not going to sacrifice the quality of the surgery."

As for my nipples, he said, they would slice off a little from the underside of the nipple and biopsy it on the spot. If it was negative for cancer, then I could keep them. But if they saw anything suspicious, he would cut them out. Afterward, my dismembered breasts—items I can still hardly bear to think about; I envision them as lumps of soft tissue, like the fibrous inside of a skinned peach, red and dripping with juices—would be sent off and sliced up further and examined for cancer, sacrificed to the gods of science. Some of the women he had operated on, Dr. Ashikari said, even women my age, learned that they already had nascent cancer inside their breasts. Later, he tells me that the literature estimates this rate at about 5 percent, and his numbers are roughly in line with that, though a bit lower—15 of his 417 patients have had this happen. The thought chills me.

Afterward, on the long drive back from Tarrytown, New York, to Edison, New Jersey, my mother and I dissected the visit. As a young nonsmoker with elastic skin and round breasts, I was a good candidate for this particular type of procedure. I liked the idea of waking up with reconstruction complete—of never having to see myself with a flat chest, as I would during traditional expander surgery, and of finishing up in one single go. Psychologically, I thought it would make things easier. Both of these surgeons had performed scores of nipple-sparing operations, mostly successful. At the appointment, Ashikari had told me that more than half of his patients maintained some feeling in one of their nipples, but that he was still trying to translate that number into hard scientific data. (Later, he tells me that about 70 percent of the patients who respond to his post-surgical questionnaire say they have some nipple sensation, but that since those are self-selected, he'd say the number is more like 50 percent.) He also pointed out that there are different sorts of sensation. Most patients seemed to respond to dull sensation—envision pushing on a nipple with a pencil eraser—not sharp sensation, delicate enough to detect a pin. So patients do not regain total feeling, he said, but some. And since the alternative option of nipple removal presented a 100 percent chance of losing all nipple sensation, I found the procedure attractive. My mother felt the importance of all this, of course, but she'd also liked how the doctors had talked to us. "I think I was relieved to see the people that would be working on your case and that they treated you with respect and that they gave you the information," she tells me years later. "It was hard to see you go to this appointment, and I think I probably held a lot of it in because I wanted to be strong for you and just be matter of fact."

Although I'd checked these guys out and they seemed top rate, in the end, it wasn't their expertise that convinced me, but my gut feeling. I liked this odd couple, and in the pit of my stomach, I felt I could trust them. If I was sailing into a strange, threatening new world, then I felt safe with Kirk and Spock at the helm.

Maybe I was not the prone patient, the damsel to their heroes, though. After all, I had sought them out and made this choice, and something about that felt empowering. During the consultation, Ashikari kept congratulating me on saving my own life. In the months to come, some people—Internet commenters, friends, and so on—would go further and call my choice brave or heroic. I don't think it was either, because on some level this didn't feel optional—I couldn't have chosen otherwise. I felt like I'd unwittingly entered a torture porn horror movie, or a Korean revenge drama, as if some psycho had broken into my house, leveled a gun at my head, and given me a choice between a slow, horrible death and mere torture. I chose pain.

11 | Ta-Ta to Tatas

I scheduled my mastectomy the same day I signed my first book contract. It felt like karmic retribution, to formalize both arrangements on the same day. One of my major life goals was to publish a book, and now I wouldn't have that as a deathbed regret. That the hospital should have called to finalize the arrangements right after the contract-containing FedEx arrived seemed ironic—the things I most and least wanted converging at the same desk, as if I had to give up something precious in exchange for my heart's desire.

When I got off the phone with the hospital, I started crying again. Putting the date on the calendar made it feel real—my sealed doom given a concrete time and location. But the day was supposed to be a happy one. I'd signed my book contract. When I mailed it at the post office, I tried to smile for the selfie of me with the Priority Mail envelope, but it came out a grimace. The Facebook congratulations I normally live for—"likes" validate my solitary existence—barely puffed up my mood. Forty-two days until someone chainsawed off my breasts.

In some ways, the countdown had begun long before that, after I decided to remove my breasts. For my last birthday with my real breasts—my twenty-eighth—I wanted lobster, something I'd heard had dropped in price thanks to the financial collapse, but the ones we bought for the celebratory dinner with my parents tasted funny, as if the crustaceans had gotten a bit long in the tooth, faintly bitter and off-flavor, the experience so unappealing that none of us would consume lobster again for years. The last Christmas with my breasts we spent at my parents' house even though it was our year to visit George's family. I remember nothing of that holiday except that a dear family friend who practiced Reiki wanted to meditate over me and I let her. We spent New Year's in Boston, but I couldn't take pleasure in the cheap champagne or the celebration. This year was going to suck. I'd gotten a giant zit on my chin right before the party and was feeling insecure about it; so I'd troweled on makeup. My unsteady hand hadn't been able to draw in the red lip liner with the precision I would have preferred, so I spent the party feeling like a sad smeared clown in a bright yellow dress, making awkward small talk and wishing desperately that I could turn back the clock.

By mid-January, in addition to committing to an always-dresses fashion statement—when you wear them with leggings, it's sort of like sporting pajamas all the time—I'd also become obsessed with painting my nails and plucking my eyebrows, ways of doing something nice for myself that didn't involve Häagen-Dazs or gin fizzes. I'd never been into the eyebrow thing for feminist reasons, but now it made me feel tidy and neat. Maybe it'd draw the eyes upward, away from my gut. I tried a new brand of makeup and used my first anti-aging eye cream. In retrospect, I now realize that with my femininity under threat from the surgery, I gravitated toward the hyper-feminine and the synthetic. If I wasn't going to be a natural woman any more, fuck it, I could wear red lipstick as much as I wanted. One of the nice and terrible things about femininity is that a lot of it's about the packaging—I could go through the motions of self-manicure, waxing, and eyelash curling whether or not my new body de-womanized me. I found my new urges peculiar because I have never considered myself a girly girl. My mother likes to joke that I entered my black phase in third grade and haven't shaken it since—even though, in her best mom voice, she usually adds that pink looks so cute on me. I'd always felt like a person first, woman second, and I viewed femininity as a constrictive lens that society superimposed over my intelligence and curiosity. Yet, evidently, I valued it on a deeper level than I'd known.

Forty-two days left with my breasts. Only forty-two. What did I need to do with them before they left me forever? Well, I wanted to have a bunch of sex with my husband, for starters, despite the libido-killing depression I was battling. Our encounters, rare though they were thanks to my emotional state, often dissolved into tears. When he touched my breasts, my mind would flash forward, and I'd envision a gloved hand gently curving a scalpel down toward my chest. This is one of the last times I'll be able to feel them, I'd think. This is one of the last times with my breasts. It made me sad. And it's not what you're supposed to think about during sex. It's not sexy; it's awful.

George was kind about this, more than kind. He hugged me and said comforting things and gave sympathy. But even this dissatisfied me because I wanted empathy, and no matter how much he tried to understand, he couldn't grasp my particular sheen of sadness—I was losing the way my breasts looked in my favorite shirt, my capacity for sexual arousal, my body parts. He couldn't understand my flavor of grief the way I wanted him to, because he is a man and he doesn't wear the cultural context of breasts under his skin every day. This frustrated me—I wanted him to share my feelings because then I wouldn't feel so alone in my grief, and though I knew it was irrational to want something from him that he couldn't give, I selfishly wanted it anyway.

My female relatives understood, but since we weren't close, they passed little notes through my mom. Several months earlier, I had written about my mastectomy decision for the *Today Show* website, an article that garnered hundreds of comments and gave me my first taste of Internet hate—assertions that I was ugly and stupid and crazy for making this choice, that I should never have children, that I'd remove my brain if I had a risk of brain cancer, that I'd never have needed to do this if I'd practiced Ayurvedic medicine, drunk water with a basic pH, and become a vegetarian, no, a vegan, no, a raw foodist, or taken some extra vitamins or a tonic that totally cured this one guy of cancer this one time. But I wasn't putting my life in the hands of anecdotal evidence—I required science-based medicine. To the one guy who told me that I should just go ahead and get cancer because it's really a fungus that can be easily treated with baking soda, to the people who wanted me to become a fruitarian yogi, I say—go pedal your snake oil and false hope elsewhere. I know what happened to Trudy, and until you can provide time-tested scientific proof published in peer-reviewed medical journals, basically, you can bite me. It's assholes like you that make things worse for cancer patients by telling them that they must have sinned greatly for God to punish them like this, as one doctor told Trudy, or that they had invited this illness into their lives with their negative attitudes. No. Nearly one in two American adults gets cancer. As columnist Molly Ivins once wrote, "I suspect that cancer doesn't give a rat's ass whether you have a positive mental attitude. It just sits in there multiplying away, whether you are admirably stoic or weeping and wailing. The only reason to have a positive mental attitude is that it makes life better. It doesn't cure cancer."

Still, after I wrote about deciding on mastectomy for the *Today Show* and after I'd gotten over the idiocy that poured into my e-mail inbox and into the comments section on the website, Lisa and Kathy both wrote to me about how emotional it made them to revisit their own surgeries in this way. My aunt Cris wanted to come be with me during the aftermath, my mom wrote to me, but I had scheduled my surgery during her long-planned vacation to Spain. It was nice to know that I had three breastless graces looking out for me, three women who had done this preventively and lived to tell the tale, lived happy, fulfilling lives even though they were still sad, who did what they had to.

The only thing worse than having a mastectomy is planning it. I began to contend with the logistical angle. I had presurgery blood tests and scans to complete, films to have sent from one hospital to another, and insurance approval to facilitate. I knew I wanted my parents there for some of my recovery, so we planned for them to come up and bring an easy chair for me to sleep in. What hotel would they stay at? When in the morning did I need to be there? What kinds of clothes could I wear afterward, and was it OK to take my asthma medicine in the lead-up to surgery?

During this time, my mother sent me two categories of e-mails—ones that focused on the surgery and the logistics around it, telling me she was proud of me for educating myself and making a difficult, informed decision, and others that carefully revolved around not-cancer, little notes about the weather or the plants in her garden, or that just said that she was thinking of me and loved me. My father isn't a sentimental guy, but he wrote me too. Since we'd both been on the gain, we decided to attempt weight loss together. Once a week, we'd hop on the scale and send our numbers, and every few days we'd send food diaries of what we'd eaten. At the time, they seemed like pragmatic, if ineffective, e-mails. In retrospect, I read them as my dad telling me he loved me, over and over again, in his lists of bad weeks full of hamburgers and good weeks full of fish.

Twenty-four days before I cut off my breasts, George cast my chest in plaster. I wanted to remember my old body, the size and shape of my slightly lopsided and completely perfect-for-me breasts. We drank beers while he dipped plaster bandages in water and smoothed them over my cellophane and Vaseline-encrusted torso. In the kitchen, we'd gone crazy with cooking—curing salmon, Canadian bacon, and our own kimchi. It struck me that we were preserving my breasts just as we preserved the pork tenderloin and the salmon filet. In a blog post about the event, I wrote, "I could feel the cool plaster molding to me, but slowly, it grew stiff and I could no longer feel his hands smoothing the edge of each strip down, just pressure on top of the carapace that had become my chest. I wondered if this is how I will feel after I recover because the operation comes with permanent loss of nerve sensation."

That same day, I sent an e-mail about my surgery to my cousin. I felt glad to be proactive about my health, I wrote to her, "but there is still some weird gap between the smart thing to do and what I emotionally want to do." I wanted to run, but I knew that no matter how far or fast I did, the crippling fear of breast cancer would follow close on my heels.

As the day drew nearer, I began to feel completely crazy. I spent three- and four-day jags in the apartment, not managing to drive George to work or even to make it out of pajamas. My mind focused like a laser on the impending misery, and I found it difficult to distract myself from the horror to come. I decided to work with my focus and try to channel it into something more positive, and I began planning a "good-bye to boobs" party, or as I privately dubbed it, "ta-ta to tatas." George and I brainstormed about the foods we could have—braised breast of veal, slippery nipples, boobtinis (with two olives, of course), prosciutto-wrapped melons, punch served in jugs. My friend Urban would use a free plane ticket to fly into Boston from Seattle for the bash. Our old roommate Chip would be there, along with four of George's good dude friends and a few of the women I'd grown close to in graduate school who worked on a literary magazine with me. We set the date for the weekend before my surgery, when I'd be most in need of distraction.

Ten days before the operation, I decided I needed to change my hair. "I am a wreck," I wrote to my lady friends. "I'm basically fine except for when I remember that my boobs are coming off, which is, you know, pretty much all the time except when I'm distracted by hanging out/watching genre TV/drinking. I needed to get that off my chest (ha ha). . . . What I need from you is hair advice." Every one of them responded, noting that I probably shouldn't change things too drastically from my usual bob, because I'd want to wake up feeling like myself. We decided that I could go asymmetrical and needed the blue streaks I'd been too timid to get in college, and I rushed off to the salon for an unduly expensive procedure that made me feel different, jagged, unbalanced, on the edge of something. I liked it. Now we could talk about my hair instead of my breasts.

Emotionally, I was laid out. I felt like I'd planned my own execution and thought of the approaching date with dread and also desire—I wanted this over with. The fear of the procedure—not just of losing my breasts but of the surgery as well—subsumed me so completely that I counted down the number of days on Facebook each morning. I reached a fever pitch. The week before the surgery, I came home drunk on the commuter rail after an unexpectedly late night out with a girlfriend who couldn't make the tata party, and I lost my phone. The next day at the mall I let missionaries talk to me for an hour. I watched three seasons of *30 Rock* in two days. I needlepointed an entire pillow, and gave myself a manicure and pedicure. George and I spatchcocked chickens, fermented batter for dosai, roasted vegetables, and cracked raw eggs over Caesar salad.

At the supermarket the morning before the good-bye to boobs party, George and I found a pair of honeydew melons to give away in the half-baked breast-pun contest we had concocted. At first we pored over the bin, searching for the ripest, most delicious ones, and came up with two of differing sizes and shapes. They didn't match. We needed melons that matched, he said, that looked white and unblemished, and suddenly, in the sinking pits of our stomachs, we knew we couldn't find what we sought in any fruit bin.

The party, which my in-laws graciously let us host at their house in Boston, helped me. I could have spent that weekend weeping, but instead our best friends kept us laughing with puns and jokes about how I was about to have perfect, perky zombie boobs. We took pictures holding the melons in front of our chests. They brought melon-flavored Midori and wheat beer (my favorite) and seven-layer taco dip and single-malt scotch and hugs and smiles and sympathy and compliments about my new hairdo and handmade greeting cards and memories of earlier, happier times when I'd been young and carefree, and assurances that those days weren't over yet either. Afterward we went out to a bar and drank beer and made plans to meet again on the other side.

And then we went home to New Jersey, so I could have my breasts cut off.

The night before, my parents take us out to what my aunt Cris and uncle Alan call "the last supper," the last meal the night before surgery. We eat some kind of Italian food and sleep in a hotel up close to the hospital, in Tarrytown, New York.

At the hospital, after we do the paperwork, I give all my jewelry and clothing to my mother, and as I hug her good-bye in my terrible hospital gown, I can see her nose getting red, and she asks the nurse one more question to delay the inevitable. This must be hard for her, to see her child do this. I want to comfort her, but I have no reserves to spare for anyone else today so I turn away. In the prep room, Ashikari and Salzberg keep things light, asking about my writing as they draw with permanent marker on my chest, marking the center of my breastbone. I feel self-conscious about my blotchy un-makeuped face as the hunky, very kind anesthesiologist hooks me to an IV tube.

I take a last trip to the bathroom with my drip, because I'm not ready. I'm not ready to lose my breasts, and I want to see them and feel their softness one last time. I open my gown in front of the mirror and try to fix them in my memory: warm, round, a little droopy. I wish I'd taken a photo of them before now, because now it's too late, and I'll forget. I'll forget them and what they looked and felt like. I will miss them so much. I want to look for longer, but I don't think I can bear up to that; so I tear my gaze away and dab ineffectually at my face with toilet paper and force myself back into the public of the OR where I know I can't break down. As they put me under, I recite some stanzas from "The Cremation of Sam McGee," a favorite poem my grandpa Gov had known by heart before he lost his wits. I barely make it through a stanza and then, I'm out.

When I regain consciousness, Ashikari and Salzberg are wheeling my bed somewhere, I think, and it seems important to try to tell them I've appreciated how nice they've been about this whole thing, how their offices had been just like spas. But I'm coming out of anesthesia, and probably incoherent. When I wake again, I feel like Frankenstein's monster, coming to life—everything seems dim and creaky, and I'm hooked to machines that beep. I can't breathe, so I call the nurse over and she allows me a puff of my asthma medication. My lungs seem made of turgid rubber, difficult to expand. She tests the oxygen level in my blood; it's fine. Probably just the pressure from the new implants, she says.

In my hospital room, George and my parents meet me. The men spent the morning making chicken soup with rice in the hotel kitchenette and have stashed a supply down the hall so I don't have to eat the hospital food. I imagine the awkward trip to the grocery store, the friendly kitchen bickering over how to do it right, shot through with the tension of waiting for a loved one to come out of surgery. It tastes clean and fresh. George doesn't want to go back to the hotel with my parents; he sleeps in a cot in my room both nights, leaving only for meals while I'm passed out.

The next day, I shuffle around the floor with my IV and my family, and a nicely dressed woman comes to our room. She's in her early fifties and thinks I'm here because I have cancer. We don't know how to tell her that I don't. She tells us about her own experience with cancer and leaves some pamphlets. In bed, I feel like I'm dying—doped up on painkillers and exhausted from the surgery and anesthesia—and all I want is for her to leave so I don't have to act polite. She is from the Reach to Recovery program that my grandmother took part in a decade or two after its inception in 1952. Survivors marched into hospitals and connected with current cancer patients without doctors present—a radical move at the time. Later, I will complain about this woman, foisting her own cancer story on me at a vulnerable moment. My mother, whose ordeal has been different, will say, I don't know, Lizzie. When you feel so horrible in the hospital, when you've had chemo, I think it can be pretty transformative to see a nicely dressed woman who has been through the same thing and know that you can make it through.

Ashikari and Salzberg come to check on me a few times, jibing at each other as they open the surgical bra, staring at the pulpy, blackened mass of my chest and nodding with approval. I'm healing well. I try not to look because I'm still in shock and I don't want to dwell on what has just happened. It's difficult to look down anyway—the surgery has moved my chest muscles around, so it's hard to bend my chin forward. My father encourages me not to be shy about using the on-demand morphine button because it's the good stuff and I shouldn't have to suffer pain. It makes my legs itch, but it also gets me so high that I don't care that my breasts have been hacked off, so I use it until it is gone.

The drive home from the hospital the following day feels interminable. Because the surgery has moved my pectoral muscles, I'm not supposed to lift anything over five pounds for the next few weeks. I'm not even supposed to push a shopping cart at the grocery. I'm too weak to close the door to the car or even fasten my seat belt. Forget opening the sticky door to our apartment.

At home, ensconced in the chair my parents brought for me, I hunger for Chinese braised duck, spurred by memories of the dishes my father and I cooked after the ham Christmas, when my mother lived in the horrible, garbage-bag-lined corner room of the hospital. Won't it be too rich for her, since she's just out of surgery? my dad wonders. Well, if she's hungry for it, my mother muses . . . Everyone bustles around the house trying to make things just right and failing, because just right would mean never having had to cut off my breasts. The duck stew tastes delicious, rich with deep soy sauce and the smooth flavor of emulsified fat. We have it over noodles with braised root vegetables, and my father simmers a vat of beef stock on the stove for the soup he'll leave us with when he returns home, though my mother will stay on for as long as we need her. Since the men are squeamish about it, my mom takes me into the bathroom to milk the blood and lymph out of the four plastic tubes that emanate from my Frankenchest. She measures the liquid that has dripped into the attached bottles and empties them, as she will twice a day for the next week or so. When she milks the drains, they create suction out of my wounds, which aches a bit, even through the painkillers I'm on.

The next morning, I wake stiff and sore, since I haven't moved in the chair all night. When I sit up, the heavy weight of my new breasts falls down and hurts. I take a painkiller and an antibiotic, and suddenly, I am throwing up. With each heave, my entire rib cage feels like it is unhinging to spill my new breasts onto the floor, as if my chest will burst and my heart will come out through my mouth and into the metal bowl, along with the pills, last night's dinner, and finally, bitter green bile. Two frantic phone calls and a trip to the pharmacy later, I have nausea medication to sustain me until I'm off the pain meds.

After my father returns home, my mother sleeps on our futon next to the chair I inhabit unless I have a follow-up appointment with the surgeons or unless she drags me out on a little excursion to get me up and around. Because she has been through so many surgeries, my mother knows exactly how to help. In between seeing Salzberg and Ashikari, we have a few hours—too short a time to drive home—so she takes me to the mall where I have my hair washed and blow-dried, something I haven't done in several days because I'm not allowed to shower until the surgical drains come out. While we are there, she finds a dress that zips up the front and insists that I try it because it is pretty and because front-closure clothing is all I can wear, since I cannot lift my arms over my head. I dress in leggings and button-down shirts that pooch out over my surgical drains and paunch. She makes George and me elaborate salads with fruit and seeds—fiber to counter the paralyzed sphincters of general anesthesia—keeps track of my pain medication, heats up washcloths so I can sponge off my limbs, and helps me remember to stretch my arms every day. At first, I can't raise them at more than a ninety-degree angle from my body, but I quickly improve. I feel better once the drains come out and Dr. Salzberg puts lotion on my hand and shows me how to massage my chest to stimulate blood flow to the skin and break up scar tissue. They have turned from black to purple to green to yellow as the bruises fade, with skin as rough and pitted as an orange peel. I am afraid to touch them, afraid to touch my own body at first, because it is sore—or even worse—numb, and I worry that I might rupture something, hurt myself without knowing it. He has to teach me that it's OK to touch myself again. Distrusting my physical self in this way feels new; I've loved and hated my body before, but I have never, ever feared it.

And at home, nearly a week after the surgery, after the general anesthesia is out of my system and I've mostly weaned myself off the pain pills, my emotions begin to return. My mother has just gone to bed on the futon, and I'm next to her on my chair in the living room, trying to sleep, when finally it hits me; my breasts are gone, and they'll never return. Once I start to think about it, I can't stop. How has my mother borne this all these years? How will I be able to bear it for the rest of my life—body parts cut out, strange appendages affixed to my chest. I will never be the same again. How did she do this with a small version of me running around? If I had a baby now, I wouldn't be able to hold her—it is still a challenge to lift my laptop. How has she carried this secret sadness for all this time? How can I bear it—how can I go on living when part of me is gone?

I wake my mother up and we talk. I cry. She holds me. And she doesn't offer false assurances that everything will be all right—she doesn't minimize or try to distract me from what I'm feeling. She tells me the truth. It doesn't ever go away, she says, the sadness. It's always there. This is a loss. But as time wears on, the pain is less keen and arrives less frequently.

Friends and family send flowers, movies, funny hats, hand-crocheted blankets, red lipstick, and the softest, pale pink zip-up sweatshirt, which I put on immediately and won't take off. The packages and well wishes make me smile and give me the activity of thank-you notes. I start a needlepoint project under my mother's tutelage. We watch endless DVDs, stretch my arms, take little walks down the block and back. And one day, like Mary Poppins drifting off into the rain clouds with her umbrella, it is time for my mother to leave. We hug gingerly—I am beginning to realize that her chest has always felt stiff to hug because her implants are firm; her breasts don't squish the way mine do, or rather, did. It will feel the same way to hug me now. And she drives back to DC in her car, leaving behind a clean apartment and a half-healed daughter and a worried son-in-law in a home that now feels empty.

It's nice to be the two of us in the house again, George and me, except that things have changed. For our entire relationship we've gone to great pains to keep things equal, and now the surgery has disrupted that balance. It began after my genetic news, an emotional asymmetry, a psychic pain that selfishly demanded attention. Now, the surgery has added a physical unevenness. I pick up some beer on the way back from a doctor's appointment and find that two six-packs are too heavy for me to carry all the way back to the car. I can't run errands for us for a while. In the supermarket, George sends me off to get pickles, and I have to abandon the jars in a display case halfway and go find him. After a few weeks, it still frustrates him that I tire so easily—I'm often too exhausted to help make dinner, or clean in tandem, as we always do, or go out in the evenings for anything other than book reporting work. We compromise—I read aloud to him while he performs household tasks, but this long-term inequity, the way this surgery has sucked the air in our relationship to me, as if my breasts are now a black hole, like the ones his astrophysicist buddies study, it changes us. It exhausts my capacity for gratitude and his for patience. I went through something and needed my way, and the chaos lasted for so long that it wore grooves into our relationship. We are ready to get back to our normal lives, but we can't—they're forever barred to us, the fruits of paradise lopped off and irreplaceable. Since we can't get back to the old normal, we flail around trying to discover the new.

I am trying to adjust to my new body. It's unsettling, to change overnight like this. I've spent twenty-eight years with my physique, figuring out what sort of necklines look good on it, balancing the top and bottom of my frame. In six months, it's changed completely—first with weight gain and now with my new larger breasts. I had asked the plastic surgeon to make them a bit bigger, not only because I heard a rumor that, depending on where a woman's nipples sit to begin with, it can make nipple placement a bit easier but also because I liked the idea of returning enhanced, and because other women listed not going bigger as a regret. But surprisingly, I've jumped two cup sizes, and I realize with a shock that, though I had only asked Salzberg to make me a little bigger, at the time I did so my boobs were at their largest ever thanks to my weight gain. I had probably jumped a cup size before the surgery, but in all the emotion, I hadn't noticed.

Some of my favorite dresses are too tight in the bust, and I have to toss them. The ones that remain, well, it's true what my larger-chested friends have been saying all along—tank tops and scoop necks that used to look pleasantly attractive now give me the appearance of a lady of the night. I'm swelled up, oozing out the sides—my cups of silicone runneth over.

At Nordstrom, where I bought my surgical bras, I let one of the staff fit me for normal lingerie. I can tell I will come back here again, even though the bras are expensive—they fit me with kindness and understanding, not just now, but right after my mastectomy, and I will never forget that. The saleswoman looks at my new breasts in the dressing room and tells me how nice they are. "More women would have mastectomies if they knew they could look like this," she says. And when the test bras come, they fit me like a glove. "This bra looks like it was built for you," she says, "as if you were the model for it." It is a nice compliment to my surgeon, I think, but it makes me feel strange—these bras were built and mass-produced for some nebulous everywoman. The individual flaws of my body have vanished, replaced by Nietzschean überbreasts.

I buy my bras tight, just like the beautiful mutant postsurgical doctor at Salzberg's office told me to. As they heal, my breasts will grow into the shape of the bras, and though at more than a month out, much of this has already happened, still, positioning them is important. The first day I wear one of these new bras starts out great, but by the afternoon I've learned that the nerves in my armpits are only mostly dead and puffy—I get sore there from the pressure of the underwire, extruding my new plastic tits forward.

In the coming months, that's what I realize—I no longer have breasts. Breasts are beautiful and natural. They belong on your wife, who will gaze down with love at your child during feedings. They are large, small, droopy, and soft, pillows to rest your head on, things bras hold up, delicate and vulnerable and sometimes lumpy. What I've got are tits. Large, perfectly formed, mass-produced tits. Aesthetically flawless, like Barbie's, but soulless and universal in their round perkiness, and downright obscene in a tank top or low-cut shirt. Tits are what your mistress has, your trophy wife, the models splayed across billboards. But the rest of my body doesn't match my twin diamonds—it's as if I put rims on my minivan—and for a long time it's confusing to grapple with this. My breasts wanted to be modestly covered, or half-displayed like a necklace in a velvet jewel case. But my tits are for flashing, for cleavage bared to the solar plexus, for unsubtle balconette bras full of artificial enhancements. If I'm going to present a lie to the world, I think, why not go all the way? I buy black lace bras, plunging necklines, and lipstick so crimson and waxy that when I kiss people hello their marked cheeks stay red for twenty-four hours. My modesty has vanished along with my natural body—after so many men and women have felt me up and fingered me on paper-covered tables, I no longer feel I have a right to it; I became promiscuous in uncovering myself to medical strangers, a prelude to the final, ultimate nakedness, beyond mere skin, where men I'd met once opened me up and changed me. Now, decency seems impossible, a grim joke, a bearded disguise so laughable that it ought not be attempted. With my curvaceous new body, even in clothing that covers me to the collarbone, I often look slutty. I can't help it—I have what still feels like bowling balls strapped to my chest. They feel as dead as bowling balls too, except for the phantom itches I can't scratch, since my skin receptors have vanished, and the tingling pain in one nipple that the surgeons say means my nerves there are healing. In crowded rooms at parties I slither past strangers, only realizing in retrospect, from their shocked glances, that I've accidentally brushed past them with my senseless chest, as if I'm driving a rental and nudged another car because I'm unsure of my dimensions. Sorry, friend, my slutty tits and I weren't trying to hit on you—I simply wanted some artichoke dip and you were in my way and I miscalculated how much space I needed. I want to show off my new assets—I want compliments about the nice new set of iBreasts I've purchased. I also want confirmation that no one knows how fake they are, unless I fess up. Since my reconstruction has made my mastectomy invisible, I also want someone to witness my pain. I flash my new rack to friends in restaurant bathrooms, shared hotel rooms, and saunas. This horrible thing happened to me, I am trying to say with my naked torso. They're beautiful, everyone answers. What else can they say?

They are beautiful. My plastic surgeon calls me "Ms. Natural," my oncologist says it's a magnificent reconstruction job. I'm impressed myself and pleased that they look so good. If I could, I'd hoist my dismembered breasts in a two-tit salute to Salzberg and Ashikari's magic hands. I couldn't have hoped for a better outcome; they look great, and I even have some nipple sensation on one side. It's just that the new rack was destined to fall short, simply because I never wanted it. I miss my old one.

Perhaps it's the skin-deep perfection of my tits that makes me so unhappy about the rest of me. I feel overripe, on the edge of rotten, bursting, like the middle-aged, overweight noir lush that our hero throws over for the girl next door with the interesting nose. My husband still loves me, though. He says so all the time. And he shows me too. But something in me can't believe him when he tells me I am still compelling to look at. He's like my mother telling me I look good in pink—after this ordeal, we're so close that his love obscures his judgment, and he is no longer impartial and trustworthy. Still, it is a relief that my husband finds me attractive even after this surgery. He loves my new body because he loves me.

It's not easy to adjust our physical relationship. After our first few encounters, when we wore masks to each other, put on brave faces—him to reassure me that I am still attractive, me to reassure him and myself that I can still be a sexual person—the emotion hits. He's touching my breasts in bed one night and I can't feel my skin, only the pressure of his fingers. My breasts used to be a source of pleasure for me; now they're numb, and it'll be like this for the rest of my life. It's a loss for him too because he loves me and because he likes to give me pleasure, and now he can't do that in the way he used to. He holds me while I sob. I think about the clinical phrase "permanent loss of skin sensation," such a throwaway line I read past on the way to surgery, a side effect that seemed like a small sacrifice compared to the certainty I'd gain from this operation. I am beginning to grasp the full breadth of its meaning. It's not the worst thing in the world; but it's still a loss, and I mourn.

12 | Heffalumpless

It's not easy to let go of fear. In the months and years that follow my surgery, I mostly feel like sunlight has broken through the clouds. No more mammograms, no more waiting for test results, no more fear. When a 2010 study comes out in the *Journal of the American Medical Association* confirming earlier results that preventive mastectomy in BRCA carriers reduces the risk of breast cancer by 90 to 95 percent, and I see the hard data that my risk is dropped into single digits, that it is now, in fact, lower than the risk of the average woman, I feel relief.

It's still unsettling that it takes more than a year to return to my normal energy levels, that I must spend months on physical therapy after a fall while reporting for my first book. While shadowing a source participating in an interactive drama in the woods—picture *Lord of the Rings* with lower production values—a monster jumps us out of nowhere, and while fleeing down the dark, rocky path, I trip. Because I know my arms cannot bear my weight—I still cannot use them to stand up from a sitting position—I roll forward over one shoulder, wrenching a back already overtaxed by adjusting to the new position of my chest muscles. I ache for months.

The knowledge of my lowered risk creates an energy void in my life. For the last twenty-eight years I have expended untold time and effort worrying about breast cancer in some form, a silent ever-present hum in the back of my mind. Now, there's no reason to fear—I've lowered my breast cancer risk as much as I can—but the absence of that emotion creates a surfeit of nervous energy. I am so used to being afraid, to organizing my life around this fear that I barely know how to behave. I am Piglet without his Heffalump, Beowulf without his Grendel, an Ahab who has slain Moby-Dick.

My Heffalumplessness means I must worry about other things—my career, bills, where George and I will move next. But I haven't worn smooth the grooves of fear, and inexorably, it seeks a new object: ovarian cancer.

As a BRCA mutation carrier, my risk of ovarian cancer is also elevated. How high? As with all the BRCA risk numbers, research is still coming in, and they're being revised, largely downward, as the families used to discover the genes typically had very penetrant genes, while many mutations we now know about may actually connote a more modest level of risk. A study of Ashkenazi Jewish mutations says the lifetime risk is 54 percent and 23 percent for BRCA1 and BRCA2 mutations, respectively. But my mutation is French and hasn't been widely studied. Another study of multiple families puts the lifetime risk at 39 percent and 11 percent for BRCA1 and BRCA2 families. A third meta-study says it's 55 percent for BRCA1 and 21 percent for BRCA2 carriers on average.

Thirty-nine percent, 55 percent for a BRCA1 mutation, what's the difference, really? No matter what way you slice it, that's a drastic, exponential increase over the typical woman's lifetime risk, which is about 1 in 72 or 1.4 percent. And the disease is a killer. Even though it's only the eighth most common cancer in women, it is the fifth leading cause of cancer-related death in women. When caught early, it's mostly curable. The problem is that only about 20 percent of ovarian cancer is detected early, according to the American Cancer Society. And the likelihood of a woman living for five years after being diagnosed with ovarian cancer is only 44 percent, a rate that has only budged by about 7 percentage points since the late 1970s. Let me say that again: an ovarian cancer diagnosis means your odds of living five more years are worse than winning a coin toss. It feels extraordinary to me that my grandmother survived the disease.

The symptoms are a hypochondriac's wet dream: bloating or feeling full, abdominal pressure or pain, pelvic pain, indigestion, constipation or peeing too much, losing your appetite or feeling full quickly, weight gain or loss, fatigue, lower back pain, nausea and heartburn, loss of energy, pain during sex, and spotting between periods. Maybe it's food poisoning, or maybe it's ovarian cancer. Maybe you've been partying too much, or maybe it's ovarian cancer. Could you be pregnant, or do you have ovarian cancer?

As it turns out, I'm not so Heffalumpless after all. In no time flat I've transferred my anxiety from breast to ovarian cancer. My period changes, and every time I menstruate for the next year, I'm thinking about my ovaries. At my age, it's vastly unlikely. Even in my family, women like my grandmother and her sister didn't get the disease until their fifties, but for a while, it still feels like it's the explanation for everything. I stub my toe. "Ovarian cancer!" I tell George. He gets a headache. "Ovarian cancer!" we shout together. We blame it for our over-seared beef, our messy house, my still-weak arm muscles.

A few months after my surgery, we grab a beer with an old friend who is now in medical school and meet his soon-to-be fiancée, an ob-gyn. I ask her whether people bug her for medical advice, and with an apologetic look in her eye, she tells me that usually, it happens in a situation like this, with a woman she's just met who is a friend of a friend, or the girlfriend of a friend, who, when the night comes to an end, confesses some lady-business problem. Two beers later, I've got my serious face on and am telling her it's probably ovarian cancer. It's not ovarian cancer, she says, it's not.

I hate feeling like a hypochondriac. As before, with the breasts I no longer have, I need reassurance that I don't have cancer, the constant call and response asserting that I don't have ovarian cancer now . . . or now . . . or now. Unfortunately, this is something medicine is not equipped to provide. To reduce your risk of ovarian cancer, you can do things like take oral contraceptives for a few years—check!—exercise—check!—be shorter than five foot eight—check!—and have children early—er . . . maybe? Of course, surgery to remove the ovaries or uterus or to close the fallopian tubes helps too, but that is dramatic to contemplate, leaving me again considering surveillance.

Like many other cancers of the internal organs, ovarian cancer is difficult to detect. The National Comprehensive Cancer Network (NCCN), a "not-for-profit alliance of 25 of the world's leading cancer centers," including participants from Stanford, Duke, Mass General, and Sloan Kettering, according to its website, recommends that BRCA carriers be screened for ovarian cancer every six months, using a transvaginal ultrasound to scan for masses and a CA-125 blood test, which looks for a certain substance in the blood produced by some ovarian cancers. Of course, as these guidelines admit, neither of those tools has been shown to be effective. Dr. Mary Daly is well aware of this fact. "We have a lot of data that there is no screening that works for ovarian cancer," she tells me. Daly is the chair of clinical genetics at Fox Chase Cancer Center in Philadelphia and chair of the NCCN's Guidelines panel on genetic/familial high-risk assessment for breast and ovarian cancer.

If ovarian screening doesn't work, I ask her, why is it recommended? Mainly because the research isn't in yet, she tells me. There are studies looking at whether taking CA-125 levels every six months, rather than every year, works better. "Frankly," she tells me, "I'm guessing those won't be successful either." This is why it is important to get high-risk women into studies looking at cancer screening, so that the data improves. And then, there is the psychological angle as well. "Some women just have to be doing something because otherwise they feel helpless about their ovarian cancer risk," she tells me. I have the sudden sense that the helplessness is not only frustrating to patients but to the doctors who want to help them as well.

Still, hope abides in the form of new research. If absolute CA-125 levels don't work as ovarian cancer screening tools, maybe more individualized ones might. A recent eleven-year study of 4,051 women used CA-125 levels differently—tracing how a woman's levels rose and fell compared to her own baseline over time. The study used the data to class 117 women into a high-risk group and gave that group transvaginal ultrasounds, which led to ten surgical operations. Of those ten surgeries, doctors found four cases of ovarian cancer—caught in early stages—two borderline cases, one case of endometrial cancer, and three benign tumors. Those are promising results but ones that are not yet conclusive. In the UK, a large-scale random screening trial of two hundred thousand women is being organized, and the results are due in 2015, according to MD Anderson Cancer Center, which performed the initial study. And that's not the only research out there, of course. Dr. Daly's group is also participating in research looking at new markers for ovarian cancer, in addition to CA-125. "So it's kind of muddy waters right now," she tells me, "but in one or two years we should have the results of this trial." I hope that both of them pan out.

In the meantime, the only thing I can do is check the last items off my ovarian cancer risk reduction list, which means carving out more organs to satiate my ravening Heffalump.

I don't feel as sentimental about my ovaries as I do about my breasts. They mean about the same to me as my spleen does—something I'd prefer to keep in, but not something that hits close to the core of who I am. That is, until I begin reading about the effects of oophorectomy. The procedure itself is far easier than what I've already gone through; it's usually performed using tiny incisions. One BRCA carrier told me, "Mine went in through my navel. No scars. And compared to a mastectomy, it's like going to the dentist."

The effects of ovarian removal, though, affect a patient long past the recovery time. Oophorectomy removes the body's main sources of estrogen and will give me instant surgical menopause, ideally, if I follow NCCN's recommendations, between the ages of thirty-five and forty, or individualized based on my family history. On a cosmetic level, as fellow BRCA1 mutant, book author, and journalist Masha Gessen wrote in *Slate* while contemplating her own surgeries in 2004, estrogen in women produces most of the physical features we associate with femininity—a narrow waist, wide hips, skin that is less red than men's, full lips, and lush hair. "Without her ovaries, a woman will often gain weight, especially around the midsection (there goes the torso), her skin will lose elasticity, her lips will lose their fullness and color, and her hair will thin (though not necessarily the same way a man's hair does). Another sexual attribute that goes out with the hormones is the libido." Removing my breasts impacted my self-image because they were part of my femininity. It feels vain and shallow to say so, but removing my ovaries feels like it will take whatever's left.

Surgical menopause does all the other things that natural menopause does and comes with risks of changes in sex drive, hot flashes, vaginal dryness, osteoporosis, and mental side effects such as loss of memory and decline in thinking skills. With natural menopause, which occurs, on average, around age fifty-one, the hormones ramp down over time, like a smoker who tries to cut down and then quit. Surgical menopause is more like going cold turkey. One day you wake up, and no more estrogen. Hormone replacement therapy can help mitigate some of the side effects, and I'll be able to have that, I hope, because I've already had a mastectomy.

And then there are the more serious side effects of early oophorectomy, the ones that are still being studied. Some studies have found links between oophorectomy before age forty-five and things like dementia, cognitive impairment, cardiovascular disease, and, most shockingly, premature death. Of course, it appears that the science is still coming in both on possible links and on whether hormone replacement therapy affects these risks—for example, one model predicted that for BRCA1 and 2 carriers, oophorectomy would increase life expectancy by some years, regardless of the use of hormone therapy. For her part, Daly suspects the premature death risk is due to the risk of developing metabolic syndrome after surgical menopause, a condition characterized by an increased risk for high blood pressure, diabetes, abdominal obesity, and high levels of fat in the blood. She tells me surgical menopause raises a woman's risk for metabolic syndrome from something like 20 percent to 30 percent. "Until recently, it wasn't really recognized that this is something you have to be aware of and start screening for a bit earlier," she tells me. But now, she says, "it makes sense to start screening these women who have early menopause whether BRCA-related or for other reasons." This means watching glucose and blood pressure levels and weight. "These are all things we can control," she reminds me, "unlike ovarian cancer that we can't control."

When I read over the list of possible symptoms of preventive ovarian removal, I can't help but feel disappointed and angry at science. The only way to lower my risk is to give up my sexual drive, my waistline, even the thickness of my hair? It feels like the quest to reduce my risk could destroy everything I like about myself. As a writer, the idea of cognitive changes fills me with fear and dread the most out of all of the possible side effects. I rely on my ability to remember things, process complex thoughts, and render them back in prose. That ability touches the core of my identity. As I told myself over and over again as a teenager, if I can't be beautiful, fine. I'd rather be smart and interesting. And what about my sexual desire? What if I lose my lust for my husband so early? That wouldn't be fair to him, and it wouldn't be fair to me either. It could change our relationship all over again, when all I want is for things to stay still and constant, to get a grip on what I have before I lose it again.

Sometimes it feels like this tiny piece of DNA will take everything from me.

But this is also hyperbole—women the world over undergo menopause, mostly natural menopause, and live to be sharp as a tack, like my ninety-three-year-old grandma who still remembers the names of all the horses on the farm she grew up on in the 1920s and 1930s, does the crossword every day, and can tell you exactly how to miter the corner on a homemade quilt with a precision her arthritic hands can no longer match. Or my mother, who still teaches children to read and manages to juggle a schedule of activities that Superman would find challenging. She and my aunt—a similarly energetic woman—have both been through surgical menopause. I hope I will be like them.

I've had it with medical journals, though, with their statistics and numbers and likelihoods of this or that. I want to know what it's like in the trenches; I want to hear what it's like from other BRCA women.

I consider starting with the women in my family, but I can't bring myself to ask my mother and aunt about their sex drives since menopause. So instead, I interview dozens of women with BRCA mutations and learn that everyone responds differently. Most women tell me it wasn't as bad as they thought it'd be, especially with hormone replacement, though it quickly becomes clear that figuring out dosage and type can take time to get right. This heartens me. One woman tells me it's the best thing that has ever happened to her. "I love being menopausal," she says. No more periods and cramps, no more cyclical mood swings or birth control. "Oh yeah, I love it," she reiterates. "I would highly recommend it to anyone."

An hour later I'm talking to a different woman on the phone. She'd been athletic and toned before her oophorectomy at age forty-three. Now, five years later, she tells me sadly, "I was at peak, and I don't think I ever will be again." Even her muscle tone changed, she says. Due to a previous bout with cancer, she cannot take hormone replacement therapy. In the wake of the operation, she tells me,

> My cholesterol has gone up, bone density has gone down. I have a lot more physical signs of aging, and my sexuality has changed a lot. My libido has plummeted. My orgasms are a lot weaker—not the same orgasmic experience at all. I feel like with the change of my sexuality a part of me is gone, a part of my identity is gone. When I look at pictures of myself from before, you know, the big before, it's a little upsetting to me. I feel aggrieved and I feel grief for the part of me that died. I think there's a bit of a code of silence there too, because not everyone has that experience.

Surgical menopause contains multitudes; it can be the best or worst thing in the world, and I won't know how it is for me until I try it.

Of course, loss of ovaries also means loss of childbearing ability. No more ovaries, no more eggs—unless you've frozen some, of course. It's a pretty final sort of birth control.

At the time of writing, I'm in my early thirties, and I think I probably do want kids. But along with the idea of fertility, the specter of ovarian removal also raises questions about whether it's ethical for me to do so. For a long time, I went back and forth on this question. On one hand, having medicalized so many other parts of my life, I am reluctant to introduce technology into what feels like the last natural lady process my body has left: conception and childbirth. There are also ethical concerns about whether BRCA is really so terrible that it needs to be wiped from the world—it's not certain that a BRCA carrier will develop cancer, and there are potential, if sucky, treatments for the condition to mitigate against these. On the other hand, the cycle could stop with me. I could ensure that the horror that has plagued my family for generations ends here, especially given that at least one study suggests that BRCA positive women are slightly more likely to have female children.

The idea of breeding a baby without BRCA entices me until I learn more about the available options. If I want to have non-BRCA biological children, as opposed to adopting, then there are two choices: preimplantation genetic diagnosis and prenatal diagnosis. The first involves in vitro fertilization (IVF). Basically, scientists combine egg and sperm from a woman and her partner outside the womb, test the embryos for BRCA when they are very small—six to eight cells big—and transfer a couple without mutations into the mother. It's expensive and comes with all the risks of IVF, like multiple births, premature delivery and low birth weight, miscarriage, ovaries that become painfully large, and ectopic pregnancy, in which the embryo implants into the fallopian tube and must be terminated. There may or may not be a slight increased risk of birth defects and of elevating the ovarian cancer risk of the mother thanks to injectable hormones used during the egg-retrieval process, which is a complicated medical procedure in itself. The certainty of having a non-BRCA baby is offset by the risk of losing the baby, birth defects, having multiples, and a possible increase in ovarian cancer risk. No thank you.

Prenatal genetic diagnosis involves getting pregnant and genetically testing the fetus, and then deciding whether to abort if it is BRCA positive. There's a 50 percent chance that I'd get pregnant with a BRCA baby, so it's possible I'd have to do this more than once. The idea of carrying a wanted pregnancy—maybe several times—to the end of the first trimester and then aborting it because it happens to carry my least-favorite gene, well, I don't think I'm emotionally equipped to do that.

If I'm able, I'd like to have my own biological children. Of course, a study has linked BRCA1 mutations to early ovarian failure, so maybe I won't be able to. But I'd like to try. It's a human desire and one I don't apologize for. In some ways, I feel that what the BRCA mutation hath given—a work ethic and a life planned to make my achievement goals early—is perhaps as much as it hath taken away.

The other thirty BRCA carriers I spoke to didn't yield much consensus on the procreation question, though everyone agreed that it's a heavy decision. Many of the men and women I spoke to had already had children by the time they tested for the gene, so their thoughts about whether they'd make that decision knowingly were purely theoretical. One man's story in particular stood out to me. He had tested positive for a BRCA2 strain linked to pancreatic cancer, and when we spoke he was in remission from this lethal disease. He had a young daughter. "I worry, what kinds of decisions does she have to make?" he told me.

> Like first of all whether to be tested or not. And do you tell your husband or your prospective husband that you have this gene that might give cancer to your kids? And do you have kids early or freeze embryos or decide not to have kids? I lament the fact that she will have to go through this.

It's hard for people of any generation to grapple with. This same man had inherited his BRCA2 mutation from his father. Shortly after he was diagnosed with the pancreatic cancer that doctors believed would quickly kill him, he went to dinner at a restaurant with his seventy-six-year-old father, his two siblings, and all their kids. His sister and brother had asked his father whether, if he had known about his BRCA mutation all those years ago, he would have had children. "He broke down crying in the restaurant based off that question, which I didn't know they were going to ask. . . . He's a very stoic person and normally would not break down whatsoever." While he wept, he explained to the family that it was a difficult question. On one hand, it would mean giving up the children sitting in front of him, but on the other hand, one of those kids had pancreatic cancer with what was at the time a terminal prognosis. As his son, who had been cancer-free for two years at the time we talked told me, "I tried to console him and say I essentially decided myself I was going to go ahead and have kids, and of course I would let my wife make the decision as well, but my glass half-full is that there will be something to fix this by the time they have to face this."

Plenty of BRCA carriers told me they would do IVF to prevent passing BRCA along—some of them had already done IVF because they had trouble conceiving or for other reasons—but many women, most of them young, told me that in the big scheme of things, they didn't think BRCA was so bad, and they put their hope in future treatments for the condition. It's hard to know what that hope might look like in practice, though, since inherited BRCA mutations, for those of us who have them, exist in every cell of the body. Joy Larsen Haidle, president-elect of the National Society of Genetic Counselors, anticipates that medical advancements will allow women to make decisions more tailored to their personal health and lifestyles—there is still data coming in on when the ovaries are best removed, for example, or whether taking just the fallopian tubes earlier in a woman's life might help protect against risk and possibly delay surgical menopause a few years. There are scientists studying how to best attack BRCA-related cancers, and whether and how to best screen for ovarian cancer. For a dose of optimism, she also pointed to major advancements in surgical techniques over the last twenty years that the world has been aware of BRCA.

All this is wonderful, of course, but it's not the magic bullet I dream of.

I'm sure some people believe it's ridiculous and immoral for me to procreate—and hey, if you don't believe in global warming, I feel that way about you too—but from my point of view, I've made my peace with it. As a society, we have only just begun to understand the predispositions to disease that lurk in our genomes. In all likelihood, every single one of us will curse our kids with some susceptibility to an illness. The difference is that mine was identified early; I am blessed and cursed to know where my Achilles' heel is and how much that risk is elevated. Of course, I would be devastated to learn a child of mind had a BRCA mutation, to know that my kid would have to endure the awful gauntlet. At the same time, this gene has shaped who I am, and I happen to like that person. I have lived a happy and fulfilling thirty-two years so far, and I'd like to live many more. Armed with knowledge of the family mutation, my kids—who would have a fifty-fifty shot of avoiding the ordeal—would have the opportunity to choose their own paths.

I am in the first generation of BRCA patients, the first generation of people who might knowingly pass this mutation to their children. With knowing, comes moral culpability—I bear it more than my mother and grandmother, who went into the process with eyes shut. A simple blood test opened mine. In this way, the knowledge itself is a burden—I am fully aware of my potential guilt and have to accept that responsibility. Is it really so bad, to have this gene? When Dr. Cook-Deegan of Duke University tells me to put on my Pollyanna hat, I do. He explains, "Basically, being a person with one of these mutations is now a predictor of not necessarily getting cancer, but of having to do something to prevent getting cancer." My kids will have a 50 percent chance of having to do something sucky, and I think I can live with that.

But in large part, it's family pathos that has pushed me into pulling this trigger—after watching my mother treated for cancer, I could hardly do otherwise. I wonder whether my potential child—let's make her a daughter since BRCA attacks women's organs more intensely—whether my dream daughter will make the same decisions. If all goes according to plan, if I do not get cancer, then she won't have to see her mother suffer, won't have to fear for her mother's life the way I did. And if she doesn't have the fear, will she have the surgery?

As Kathy puts it to me in an e-mail about her daughter Ellie, "Both you and I experienced our mothers constantly being treated/watched for cancer. Pretty intense. Ellie hasn't, thank goodness, had that experience. However, cancer is not as real for her either." It's then I realize: if I have kids and I'm not one of the few unlucky people who get cancer even after having preventive surgery, then they'll be in Ellie's shoes. So I call her.

Growing up, Ellie tells me, "I probably thought more about heart disease than cancer," because both of her grandfathers suffered from heart problems and because her mother shielded her from knowledge of the family plague. Ellie tells me she didn't find out about it until she was twenty-two, though her mother and brother remember at least one family talk about it before then, one that apparently didn't sink in. "Obviously, I knew about my grandmother [El] and obviously that there was a lot of history in our family, but I definitely had no idea about the extent and definitely had no idea about my mom or aunt," Ellie says. She is five years my junior and learned of the family BRCA mutation the same year I received my test results. Her mother's preventive mastectomy shocked her, in part because for her undergrad thesis in psychology she'd helped with a research project on BRCA, learning about the condition and taking down the histories of many women in our shoes. "The entire time I had no idea it was so part of my family, so when my mom first told me it was really a shock," she explains. And yet, she understands why her mother decided not to discuss the issue with her. She believes the knowledge would have burdened her growing up, until she was old enough to process it. For Ellie, the family legacy "seemed more of an abstract idea or concern" for the first few years after she learned of it. In her early twenties, "I was working like eighty hours a week, ninety hours a week, trying to keep my head above water, and I think that made me push it into the back of my mind because I had so little energy for it," she tells me.

The news changed Ellie's relationship with her mother. It felt weird "for me to not know that my mom had done all that and all this has been weighing on her for twenty-two, twenty-three years and she never said anything. . . . I guess it made us closer in some ways. I feel like I understood her life better; I understood what she went through in her twenties. . . . I could just see my mom more clearly now, as the twenty-something year old, her mom having just died, having this significant concern and being single." The news made her mother "more human, for lack of a better phrase, especially as I'm now this age and depending on the day I do or don't know what I want to do with my life."

Like me, Ellie sat with the knowledge of her mother's positive genetic test for a few years. Over those years, her mother encouraged her to get tested by her mid-twenties, and they talked over all the options together. She also opened up about her doubts and fears to her boyfriend of one year. "He was the first person I ever talked to about it outside of my family," she says. "And he's been very supportive in terms of helping me talk it out about the effects of getting tested, let alone afterwards." Ellie describes herself as the sort of person who always wants to know "the most information about everything as soon as possible," so she surprised herself—and her boyfriend—by vacillating about the test. Eventually, in 2013, at age twenty-six, she went to her ob-gyn's office and had blood drawn—her doctor felt that she was so informed about the options and risks that she didn't need to go to a genetic counselor—and the simple act of getting tested made her feel immediately better. "I felt relief that one way or the other I would be able to make a decision," she says.

While she waited for the test to come back, her grandmother—her father's mother—died. So rather than feeling anxiety about the test, "I was upset about and processing that." For a moment, on the phone, I flash back to my own grandmother's funeral, a mere six weeks after my own test results.

But my story is not Ellie's story. She felt relieved when she received her results, she tells me, "but even then it took me a month to realize what that meant, . . . that I wouldn't need to be so concerned about it and all of the 'if this happens,' 'when this happens,' 'when I do this,'—I didn't have to do that anymore." When she tells me about her negative results, the flash of pure joy stays with me all day. Someone has escaped. One of us has escaped, and she doesn't have to continue down the horrific decision tree, doesn't have to eat at the BRCA restaurant where all they serve is gristle soufflé and castor oil soup—she is free. The coin landed right side up at last.

Still, she must have prepared for the worst, so I ask her if she knows what she would have done if the results had been otherwise. "Not with certainty," she says, "but I think I probably would have gotten surgery. Kind of for the same reason—I wouldn't want to spend my life waiting for a shoe to drop."

If I have children, I hope they will be like Ellie—smart people engaged in the world who thoroughly think through decisions, even the decision about whether to test at all, with the help of people who love them. And yet, BRCA has shaped her life too, has changed her relationship with her mother, singing its silent pedal tone of fear and uncertainty for a few years until she decided to make her move. It's not an easy or simple thing to deal with. A BRCA mutation denies closure. One surgery may not be enough for me—I must contemplate another. My mother still feels guilty; it hits her in fits and starts. Occasionally she still cries about it, calls to apologize and say she wishes I didn't have this gene and that she's sorry I do. I don't mind so very much, I tell her. After all, this gene has made me who I am. There are worse things in the world.

13 | Through the Looking Glass

What I am living with is just one possible outcome of a BRCA1 mutation.

In December 2012, when I am partway through writing this book, I get a call from my friend Sarah. She and I met in graduate school, and we've been through a lot together—starting the literary magazine *Fringe*, marriage, her divorce, my mastectomy. Though we live in different states now, we still meet up occasionally to watch our favorite TV shows and cook together. When she calls me, shortly after Christmas, I can immediately tell that something is wrong. It's Cheri, she tells me.

Sarah and I are very close—but Cheri and Sarah are so close that their friendship transcends the word. They have been best friends since junior high and are essentially family at this point. I know and like Cheri. We're in each other's orbit—she worked on *Fringe* for a while, I've been to dinner parties at her place, and she and Sarah once drove eight hours to play a role-playing game with me for my first book. We run into each other at Sarah's parties, and I usually ask after her when Sarah and I talk. She's the best friend of one of my best friends. And as it turns out, unhappily, we have far more in common than that.

That December, Sarah has called me in a panic to tell me that Cheri has breast cancer. I feel her fear palpably rolling through the cellular space and into my body. I'm shocked, but not as shocked as I am when Sarah calls me again after New Year's. It's not just breast cancer, it is Stage IV breast cancer that has already metastasized to Cheri's lungs, lymph nodes, and liver—incurable cancer that is grade 3 on a scale from 1 to 3, which means the cells look abnormal under a microscrope and are dividing very quickly. It's also triple-negative, meaning that hormone therapy won't help. Cheri's cancer is advanced, aggressive, and will ultimately be lethal. And it is caused by a BRCA1 mutation she didn't know she carried.

Five months later, after Cheri has been through biopsies and radiation and two rounds of ineffective chemo, she and Sarah go on their long-planned trip to Europe with Cheri's sister and stepmother. It was important to everyone that they still go despite the dire news. It's a difficult trip—not just because of the diagnosis, but because it is Cheri's first time away from her infant son since his birth. As the trip progresses, Cheri notices that half her face won't smile right in the photos, so she e-mails her nurse. At the Louvre, she and Sarah split off to see the winged Nike of Samothrace together, a white marble statue of the goddess of victory, heavy drapery pressed against her breasts, wings arched back in triumph. They pose for pictures and wander into the Egyptian wing of the exhibit hall together and are looking at sarcophagi when the phone call from the doctor comes, along with the news that either it is a palsy or the cancer has spread into Cheri's brain. But she can finish the trip and see the doctor first thing when she returns. Cheri tells me the news was "just enough to give me a sliver of hope but be worried the rest of the trip, because deep inside I know the truth."

When Sarah tells me about the sarcophagi, how they reeled among the elaborate coffins in the exhibit hall, all I can think about are Imhotep's words in the Smith papyrus, the simple statement that there is no cure.

When they return from Europe, the MRI shows many small inoperable tumors in her brain. It can be hard to treat brain tumors, because much chemo cannot break the blood brain barrier. Cheri, age thirty-four, undergoes whole-brain radiation.

The diagnosis ripples outward through my circle of friends. Within less than a year, Sarah has remarried, to a divorcé, and is stepmother to his four kids, her life accelerated, perhaps, by the new knowledge that time is limited and precious. With the exception of Sarah, none of us are super close to Cheri, and we feel as if we are not entitled to the scope and intensity of our grief. When we meet up, we talk about how much this has affected all of us—we weep at home after reading her husband's updates on the private site dedicated to her care or after she posts a new blog entry about living with this diagnosis. Two of our friends meet up with Sarah to try out a new bar in Boston, and they all begin crying so hard that they joke they can never go back there. The *Fringe* editors who knew her—all women, all close to Cheri's age—band together to do something for her. Though we exchange many e-mails about what we should buy her, we feel thwarted by the cancer too, by our inability to help in a meaningful way. We pay for someone to clean her house. It feels like a pitiful offering.

It is unendurable. It is miserable. It is beyond crappy. And we are all distanced from the epicenter, where the vast blue space of hope must be limited, folded up over and over again until it is small enough to swallow. A long, healthy life for Cheri is unlikely, so we hope she will live a good life for as long as possible, even while understanding that her cancer did not respond to the first two courses of chemotherapy, and now it is in her brain.

The fall after her diagnosis—nearly a year later—I decide to visit Cheri and interview her, if she is willing. It's hard to schedule our meeting, since first she is recovering from whole-brain radiation. "I had that for fourteen days," she tells me. "That sucked." Then she is in the middle of a chemo cycle, which means she is tired for nearly a whole week after she receives treatment. But eventually, we find a day and time, and although I'm terrified that I'm getting a cold—not so bad for me, but dangerous for a person whose immune system is weakened by chemo—we settle on a meeting.

The truth, though, is that I am scared to see her. As we rattle forth about dates and times, as I make the four-hour drive from New Jersey up to her house in Massachusetts, I revisit scenes from my childhood.

After all her own illnesses, my mother has a soft heart for other people in their times of need. Growing up, it felt like she was always baking crumbed chicken pieces to take to families with people who were sick, dying, dead. She took women from her church to Nordstrom to be fitted for prostheses and wigs, served as a member of the prayer chain, remembering people on the list before bed. She seemed physically sick when one of her church friend's children got asthma, like I had, or a rare blood disorder, which I didn't. Sometimes she'd babysit other people's kids for a few hours to give the adults time to sort through the insane hash of their lives.

I often played with those kids. For a few weeks in kindergarten, another girl came over a couple of mornings a week. We'd all heard her mother had died, and we whispered about it on the school bus, shrinking away from her as if she might contaminate us. In fourth grade, I became friends with a girl whose mother developed breast cancer, so my mom scheduled twice-weekly playdates that lasted until the troubled girl, jealous that I had advanced further in ballet, nicked my toe shoes. A few years later, her mother died.

Most of all, I remember a woman from our church who developed terminal breast cancer when I was in high school. Her kids were a few years older than me, and I knew them from youth group—though they were way cooler than I was, so we weren't friends. Our moms, both active in the church, had a very friendly relationship, though. After hospice had been called in, I went with my mother to deliver a large batch of my father's beef soup, homemade down to the hand-diced meat and meticulously skimmed stock. I remember stepping over the knotty roots of a tree outside their house with my tray of goodies. Inside, the absence of the teens relieved me—I was an awkward kid. Nothing could have prepared me for the sight of this woman. Her skin stretched so tight over her skull that she looked like the walking image of death. Her skeletal arms reached up to touch my face, the fingers pinching my cheek, and she laughed hysterically as she told us that the cancer had spread to bone and brain. Her husband looked on calmly and tried to reassure her.

My mother has never been one to flinch at the body, to deny the lived experience of anyone; she sees the essential humanity and goodness in people. She kept up the light patter of small talk for fifteen minutes, and then made sure to show herself out over our hosts' protests, with the firmness of someone who knows how tiring visitors can be after chemotherapy.

That woman, her desperation and her suffering, frightened me. My life became a waking prayer for her as I walked the high school halls, and before bed: "Please don't let her die, God," was all I could think. Two weeks later, when she died, I couldn't bring myself to go to the funeral. Eventually, I wrote an essay about her during my senior year, titled "Atheist."

Cheri greets me with a hug at the door. She looks robust but tired to me, and most of all, she still looks like Cheri. She's covered her head with a pretty silk scarf that looks fashionable, like she is some sort of European doyenne in hiding.

I am surprised by how normal it feels, to hang out with Cheri and gossip about mutual friends and talk about cancer treatment. I soon get over myself—it has been unfair of me to put the burden of my relationship with cancer onto her. She is still here, living with metastatic cancer rather than dying from it, though she is very matter of fact about her illness and the way her treatment makes her feel. She has pragmatically accepted her diagnosis and is willing to talk about it and how she got to this place.

Almost two years before, she had given birth to her first child, a baby boy, whom she breast-fed. A month later, she noticed a lump on her upper left sternum. Her husband thought it felt like a tight muscle. She assumed it was related to breast-feeding and waited a few months before calling her midwife, who thought it was nothing but ordered an ultrasound anyway. The ultrasound came back negative, and the midwife scheduled Cheri for follow-up in a year. Cheri wanted a second opinion. Her primary care physician wasn't concerned about the lump but suggested she see a surgeon just in case. The surgeon ordered another ultrasound that came back normal and followed up with Cheri a few months later, when he ordered a fine-needle biopsy, where they use a small gauge needle that the doctor—by feel—inserts into the tumor to withdraw cells for testing. Of course, fine-needle biopsies do have a false negative rate of about 5 percent. Her results came back negative. A few months later, same drill—another ultrasound, another negative result. By now nearly a year had passed, with her doctors chalking up the lump to hormones related to breastfeeding. With her son almost a year old, Cheri began weaning him. Four months later, she visited her doctor again. The lump had grown a bit. So Cheri had another ultrasound. The doctor called her at home on Friday night to explain that they had found something. On Monday, Christmas Eve, Cheri had a surgical biopsy. She got the phone call two days after the holiday. "I just knew it was going to be bad," she told me, "because it had been so long" since she'd discovered the lump. On New Year's Eve she woke up from surgery to lodge a port for chemo and blood draws beneath her skin to the news that it was stage IV and had spread to lungs, liver, and the lymph nodes in her chest. Cheri said she and her husband were simply "out of our minds." The chain of events that led her to be diagnosed with the most advanced, hardest to treat, and aggressive form of cancer out there boggles the mind. "It was like the perfect storm, basically," she said.

Because she was a young, premenopausal woman with triple-negative cancer, doctors immediately tested her for a BRCA gene and discovered she carries a BRCA1 mutation, just like me. She's not sure which side of the family it comes from but thinks her mother's side is more likely, since it has five cases of breast cancer in it, including two great-aunts, two great-grandmothers, and one distant cousin, although most of these cancers occurred around the age of menopause. One of her mother's three brothers developed prostate cancer, which could be related. Her father's side has mostly men in it, although her father has one cousin who developed breast cancer young. Either her father is a silent carrier of the gene, or her mother is a carrier who has beaten the odds and has not yet developed breast or ovarian cancer. She may never know which—her mother is not in a financial position to test. Her sister, who is a half-sister, has tested and is gene-free.

With the cases of cancer in her family relatively distant—Cheri does not have a first-degree relative (a parent, child, or sibling) with breast cancer—she was not offered testing as a young woman. She even recalled one doctor telling her that because she didn't have any first-degree relatives with cancer, the family history wasn't really a family history—in the absence of an affected aunt or sister or mother, it didn't really count. Her relatives had always said that cancer ran in the family, and that was what she told her primary care physician, who said that, in deference to that fact, they'd start screening her early, like at thirty-five. "Ironically," she tells me at our meeting, "I turn thirty-five next week." She'd never even known that women could get cancer so young, she says, until she heard about my mother around the time of my mastectomy, "never knowing that . . . that could be me." She wishes she'd known; she wishes she'd gotten tested. Her particular mutation is in the 87 percent lifetime risk range. She sighs as we talk about all these ifs—if she had just seen a different doctor, if she'd had a genetic test, if she had had a more accurate biopsy procedure, if, if, if. It's possible to drive yourself crazy with ifs, she says. But Monday morning quarterbacking doesn't do any good. She has to live with the reality of now.

She has a husband and a kid and wants to stay with them as long as she can. She's not interested in knowing her prognosis though and had her therapist call her oncologist to tell the doctor to stop pushing Cheri to talk about it. "I was like, you know, if you tell me I have three months to live, I'm going to give up. I feel that's how I feel. There's no point to talk about it until it's really the end. Until you have weeks left to live."

Amid a sea of pink ribbons, Cheri tells me, "metastatic breast cancer is kind of like the shady underbelly that nobody really talks about." To me, it's also the other, darker side of BRCA, given the aggressive strains of cancer bred by many mutations and given the disproportionate cancer burden shared by this community. And the truth is, women living with metastatic breast cancer don't have many options. According to a 2010 study, only 2.5 percent of the American Cancer Society's 2009 funding went into developing treatments for metastasis, while numbers for other international cancer groups, such as German Cancer Aid or the UK's National Cancer Research Institute in the mid-oughts remained similarly low (4.3 and 5.1 percent, respectively). A 2004 look at the National Cancer Institute's grants since the 1970s in *Fortune* magazine found that less than half a percent of proposals focused on metastatic cancer. Meanwhile, up to 30 percent of patients who develop early-stage breast cancer will eventually end up with metastasis. According to the National Cancer Institute, about 5 percent of all breast cancer cases are metastatic. The amount of funding that metastatic breast cancer receives is not proportional to the number of patients who need treatment.

"It's like they've given up. It's like, 'They're dying, so why should we fund research projects for that,'" Cheri says. Looking into her eyes, I have an inkling of what it would mean to have a treatment, at least some sort of chemotherapy that actually worked to combat her cancer and extend her life. For Cheri, living an extra year could mean seeing her son turn four or five or six, or better yet—even older.

Between Cheri and myself, there's no question of who got the shorter end of the stick. The struggle Cheri faces is far more than anything I have faced, or hopefully will ever face. In a just world, there would be no metastatic breast cancer diagnoses on Christmas or New Year's or ever.

Still, I have some feeling that we are two sides of the same coin, two different Alices trapped on different sides of the BRCA looking glass. We're both writers. She's writing a journal for her son, so that he will be able to get to know her and feel connected to her. It's hard to write, she says, because she wants to protect him from what she's going through, to explain her life outside all this, her life as a young woman. "I don't want it to be about sadness," she says. And should she write it for the eight-year-old him or the twenty-year-old him?

She is young, like my mother, with a husband and a young child, like my mother had, but they are living the nightmare. We have the same genetic mutation. I had the family history to warn me. She did not. I am heartbroken that the medical advance that allowed me to reduce my cancer risk could not benefit her, and suddenly, the sacrifice of my breasts doesn't seem so costly.

After I have written about Cheri like this, and we have e-mailed back and forth about my words, things continue to go badly. In the spring of 2014, less than two years after her diagnosis, she enters hospice care for a few short weeks. But you already know the ending of this story. She was only thirty-five, and will never be any older.

I am lucky. I am lucky to have lots of women in my family, enough to show the clear pattern of breast and ovarian cancer. I also hold many sorts of privilege. I'm a white woman. In the context of BRCA, this has immediate repercussions. African American people are far more likely to receive uncertain results on their genetic tests than any other ethnicity. A full 16.5 percent of African American patients who test for a BRCA mutation receive ambiguous results, according to the National Cancer Institute. Whether this is the result of some fluke of biology—perhaps the mutations in question are unusually rare and therefore harder to study, or perhaps the high rates of cancer in such families are part of the interaction between several genes rather than the result of just one and are thus harder to puzzle out—or institutionalized racism, I cannot say. Dr. Daly suggests it may be a combination. She points out that the number of unknown findings in African Americans has reduced as more of them get tested and more data becomes available. To me, this suggests that access to medicine and to genetic testing may be an issue in getting black families to test and in accomplishing the needed research. I think about how Skolnick did his work on Mormons because they had such excellent genealogical records, and my thoughts jump to the legacy of slavery, which ripped people out of their cultures and dispersed them across the United States. Daly also suspects that there may be some biological differences, on average, to explain why African American women tend to get breast cancer at younger ages than Caucasian women and tend to have more aggressive forms of the disease, including an increased likelihood of developing triple-negative breast cancer—perhaps the mutations in this population are simply more rare and less frequent, rendering them harder to study. Whatever the cause, my whiteness made it more likely that the genetic mutation I carry would be identified.

I'm also middle class. I had health insurance, which many Americans in 2010—the year of my mastectomy—did not. I could afford to test and to have surgery, aided by my parents, who are well-off enough to have helped with some of the co-pay, a fraction of the $20,000 or so my mastectomy and reconstruction cost my insurance provider. Wealth can't buy you love or happiness, but it sure can make a sucky experience a million times easier and more comfortable. I didn't have to sell my house, as one BRCA carrier I spoke with did, to afford surgery I saw as essential to my health and well-being. I also had the time and money to shop around and find surgeons I truly trusted, which may have given me access to a higher level of healthcare than the average patient.

I was born female and identify as a woman, which has colored my interactions with the medical system as well. Because I am a cis woman, there is substantial data to guide my decisions as a BRCA patient. In contrast, when I searched the PubMed database of over 22 million scientific papers for the terms "trans man and BRCA" and "trans woman and BRCA" I netted a grand total of two papers, neither of which appeared to examine the unique issues of trans individuals with BRCA, many of whom take hormones that might presumably affect their risk of developing breast and ovarian cancer.

That issues around BRCA deal with body parts associated with femininity—breasts and ovaries—also raises subtler issues around how trans patients interface with the medical system. Would a doctor remember to ask a trans man with a BRCA mutation who has retained his ovaries about cancer surveillance? What about trans women and mammography? At the same time, for such patients to ask their doctors about the body parts associated with BRCA could be a psychologically fraught interaction. I can imagine that a trans man invested in his gender presentation might feel complexly about asking his doctor for an ovarian ultrasound, for example. I don't pretend I understand the intricacies of seeking medical care as a trans person—or indeed, as anything other than a straight, middle-class white cis woman—but I do understand that the journey of the several trans patients I encountered is different, and probably more difficult, than mine.

Finally, I'm educated, which made navigating the statistics and strange vocabulary around this rare condition much simpler than it would have been otherwise. It may also have enabled doctors to see and respond to my concerns in a different way than they otherwise would have.

I've undoubtedly missed a few elements of my privileged existence that have eased the process in ways I haven't yet considered, and my narrative here is told through the lens of that privilege. The diagnosis and surgery weren't easy, but my position in the world made them easier. Though for reasons of space and familiarity I have focused on my family's experience, my story is far from the only BRCA story—but it is the only story that is explicitly mine to tell. Different perspectives—both in the discourse and research around BRCA—would benefit the community of BRCA carriers greatly. I hope, but don't hold my breath, for a world in which the advantages I enjoy—of having researchers study my condition, of affording and having access to my choice of medical care, and of education that empowered me to seek and understand treatment—are available to everyone.

BRCA continues to affect my family. In the winter of 2013, my cousin Lisa, who had her mastectomy in the early 1980s, finally tested for the family's mutation for the sake of her daughter and son. Her results came back negative. As she wrote to me, "I am relieved in many ways, mostly because my daughter and son are cleared. However, it is odd because, like you, it's been such a part of my life forever. In some ways, and this sounds odd but I think you'll understand, I'm no longer special in that way. No longer part of that girl's club. The negative result has made me rethink my health, the dynamics of my family of origin, my destiny, and my legacy. I've had three surgeries (the original mastectomy and implants, the removal of the implants, the second replacement), which made sense based on all of the information I had at the time. In some ways, one of the reasons I put it off for so long was the fear that I would be negative and had gone through all that for nothing."

Lisa may not have had the gene, but she shared in the sisterhood of fear. If she had been able to test as a young woman and had tested negative, she wouldn't have had the surgeries, she tells me. But like all of us, she did her best with the knowledge she had at the time. And now, the specter of the disease that killed her mother, ovarian cancer, no longer hangs over her as it once did. She wrote, "I realize I'm not immune from getting ovarian cancer and that my odds are the same as anyone. But it is good to know my odds are normal. That I'm just normal."

I, on the other hand, am not normal, and am leaning toward an oophorectomy sometime in my late thirties or early forties, though

I dread the procedure and all that comes with it. Until then, I will muddle through the best I can, loving my parents, laughing about robot jokes with George, and perhaps having children with him. The thing about these BRCA issues is that they never go away. My mother still calls from time to time, saying that it hit her hard that day and she had to cry a bit. Sometimes it hits me too—I am amazed to have any tears left to shed over the scope of my family's ordeal or over my own missing breasts. As my cousin Kathy told me when I worried to her about oophorectomies or feeling whole again after my mastectomy or passing on a BRCA mutation to another generation, "It isn't there all the time. It fades over time. Because if you think about it, this is the time when it's the most relevant and intense for you, right now for you in your life." I like the idea that these emotions will dim, but I don't want them gone entirely—they're part of who I am; the rush to beat an early death I saw as inevitable has influenced the core of my identity, reminding me not to waste time and to stay busy accomplishing those things I'd regret if I died tomorrow. Perhaps this explains why my mother can't sit still—there are too many wonderful projects left to embark upon. I intend to take her example to heart.

Although my BRCA mutation has deeply affected my body, my emotions, and my sense of self, it does not define me. I leave that to my writing, to the friendships I have formed, to the leftover stew in my kitchen, or my hoard of homemade pickles, and more than I'd like to admit, perhaps, to my collection of funny hats.

Acknowledgments

I am so grateful to have such a wonderful and supportive extended family on both sides; without their kindness, generosity, and openness, I could never have written this book. Thank you to my parents, Dick and Gretchen, and to my husband, George, for always being there for me, no matter what; to my three Amazonian graces, Cris, Lisa, and Kathy, for their kindness during my surgery and their fortitude in reliving their own experiences for me during interviews; and to my uncles Mark and Alan, my second cousin Ellie, and my grandpa Roy for sharing memories and emotions about the family legacy as well. I'd also like to thank the world's best in-laws, Cathe and David, for their support. I am indebted to Cheri for opening her story to me at such a difficult time.

I am grateful to the organization FORCE, both to the thirty-odd women and men from that site who shared their stories with me, and to the many forum participants who eased my mind before and after my surgery. Dr. Robert Cook-Deegan, Dr. Mary Daly and NCCN, Dr. Otis Brawley and the American Cancer Society, Sir Mike Stratton, Joy Larsen Haidle, and many other sources gave generously of their time and energy.

Thank you to Vidya Rao, who first published my writing about BRCA at Today.com, and to my agent Jane Dystel for suggesting there might be a book here. Thank you to my editors Cynthia Sherry and Michelle Williams, as well as to Ewurama Ewusi-Mensah, Ellen Hornor, and Mary Kravenas of Chicago Review Press for their work. Thank you to Eleanor Saitta for a transformative phone conversation, and to my father for assisting with legal research around the Supreme Court case and for serving as my first reader. I'm deeply indebted to the Corporation of Yaddo, for providing me space, time, and pleasant companions while I completed a substantial portion of this manuscript. Jerry Williams, John Kelly, and Samuel Freedman provided valuable guidance, and my fellow writers Chip Cheek, Urban Waite, Boomer Pinches, Shannon Ward, Toni Kan, and Tatiana Ryckman kept me going when times got tough.

Thank you to Janell Sims, Julia Henderson, and Sarah Miles, and to the many others who listened to my worries or had a beer with me when it all became too much.

Selected Bibliography

Major sources of research are listed here for the interested reader. Full citations can be found in the notes, and a full bibliography of all works cited can be found at www.lizziestark.com.

American Journal of Cancer. "Joseph Cold Bloodgood: November 1, 1867–October 22, 1935." 26 (1936): 397–98.

Antoniou, A., P. D. P. Pharoah, S. Narod, H. A. Risch, J. E. Eyfjord, J. L. Hopper, N. Loman, et al. "Average Risks of Breast and Ovarian Cancer Associated with BRCA1 or BRCA2 Mutations Detected in Case Series Unselected for Family History: A Combined Analysis of 22 Studies." *American Journal of Human Genetics* 72, no. 5 (May 2003): 1117–30.

Armstrong, K., J. S. Schwartz, T. Randall, S. C. Rubin, and B. Weber. "Hormone Replacement Therapy and Life Expectancy after Prophylactic Oophorectomy in Women with BRCA1/2 Mutations: A Decision Analysis." *Journal of Clinical Oncology* 22, no. 6 (March 15, 2004).

Aronowitz, Robert. *Unnatural History: Breast Cancer and American Society.* New York: Cambridge University Press, 2007.

Bartlett, Willard. "An Anatomic Substitute for the Female Breast." *Annals of Surgery* (Philadelphia) 66 (1917): 208–11.

Beattie, M. S., B. Crawford, F. Lin, E. Vittinghoff, and J. Ziegler. "Uptake, Time Course, and Predictors of Risk-Reducing Surgeries in BRCA Carriers." *Genetic Testing and Molecular Biomarkers* 13, no. 1 (February 2009): 51–56. doi: 10.1089/gtmb/2008.0067.

Berg, Wendy A., Z. Zhang, D. Lehrer, R. A. Jong, E. D. Pisano, R. G. Barr, and M. Böhm-Vélez, et al. "Detection of Breast Cancer with Addition of Annual Screening Ultrasound or a Single Screening MRI to Mammography in Women with Elevated Breast Cancer Risk," *Journal of the American Medical Association* 307, no. 13 (April 4, 2012).

Bleyer, Archie, and Gilbert Welch. "Effect of Three Decades of Screening Mammography on Breast Cancer Incidence." *New England Journal of Medicine*, (November 22, 2012), 367: 1998–2005.

Brewster, Abenaa M., and Patricia Parker. "Current Knowledge on Contralateral Prophylactic Mastectomy Among Women with Sporadic Breast Cancer." *Oncologist*, 16(7), (July 2011): 935–941.

Brinker, Nancy G. with Joni Rodgers. *Promise Me: How a Sister's Love Launched the Global Movement to End Breast Cancer.* New York: Crown Archetype, 2010.

Byrne, John A. *Informed Consent: A Story of Personal Tragedy and Corporate Betrayal . . . Inside the Silicone Breast Implant Crisis.* McGraw Hill, 1997.

Calderon-Margalit, R., and O. Paltiel. "Prevention of Breast Cancer in Women Who Carry BRCA1 or BRCA2 Mutations: A Critical Review of the Literature." *International Journal of Cancer* 112, no. 3 (November 10, 2004): 357–64.

Cann, Rebecca L., Mark Stoneking, and Allan C. Wilson. "Mitochondrial DNA and Human Evolution." *Nature*, January 1, 1987: 31–36. doi: 10.1038/325031a0.

Castanias, Gregory A. "Brief for Respondents." AMP et al., v. Myriad Genetics et al., No. 12-398, March 7, 2013, 32.

Chen, Sining, and Giovanni Parmigiani. "Meta-Analysis of BRCA1 and BRCA2 Penetrance." *Journal of Clinical Oncology* 25 (2007): 1329–33.

Committee on Health, Education, and Pensions. Report on Genetic Information Nondiscrimination Act of 2007. 110th Congress S. Rep. 110-48, 110th Cong., 1st Sess., (April 10. 2007).

Daly, Mary B., Robert Pilarski, Jennifer E. Axilbund, Saundra S. Buys, Beth Crawford, Susan Friedman, Judy E. Garber, et al. "Genetic/Familial High-Risk Assessment: Breast and Ovarian." NCCN Clinical Practice Guidelines in Oncology (NCCN Guidelines ®, version 1.2014). National Comprehensive Cancer Network. Accessed March 7, 2014. http://www.nccn.org/professionals/physician_gls/pdf/genetics_screening.pdf.

Davies, Kevin, and Michael White. *Breakthrough: The Race to Find the Breast Cancer Gene.* New York: John Wiley & Sons, Inc., 1995.

de Cholnoky, Tibor. "Augmentation Mammaplasty: Survey of Complications in 10,941 Patients by 265 Surgeons." *Plastic and Reconstructive Surgery* 45, no. 6 (1970): 573–77.

Domchek, S. M., et al. "Association of Risk-Reducing Surgery in BRCA1 or BRCA2 Mutation Carriers with Cancer Risk and Mortality." *Journal of the American Medical Association* 304, no. 9 (2010): 967–75.

Evans, D. G. R., A. D. Baildam, E. Anderson, A. Brain, A. Shenton, H. F. Vasen, and D. Eccles, et al. "Risk Reducing Mastectomy: Outcomes in 10 European Centres." *Journal of Medical Genetics* 46, no. 4 (April 2009): 254–58.

Finch, A., M. Beiner, J. Lubinski, H. T. Lynch, P. Moller, B. Rosen, J. Murphy, et al. "Salpingo-oophorectomy and the Risk of Ovarian, Fallopian Tube, and Peritoneal Cancers in Women with a BRCA1 or BRCA2 Mutation." *Journal of the American Medical Association* 296, no. 2 (2006): 185–92.

Fisher, B., J. P. Constantino, D. L. Wickerham, C. K. Redmond, M. Kavanah, W. M. Cronin, V. Vogel, et al. "Tamoxifen for Prevention of Breast Cancer: Report of the National Surgical Adjuvant Breast and Bowel Project P-1 Study." *Journal of the National Cancer Institute* 90, no. 18 (1998): 1371–88.

Fulda, K. G., and K. Lykens. "Ethical Issues in Predictive Genetic Testing: A Public Health Perspective." *Journal of Medical Ethics* 32, no. 3, March 2006: 143–47.

Garcia, C., J. Wendt, L. Lyon, J. Jones, R. D. Littell, M. A. Armstrong, T. Raine-Bennett, et al. "Risk Management Options Elected by Women after Testing Positive for a BRCA Mutation." *Gynecologic Oncology* 132, no. 2 (February 2014): 428–33. Accessed March 6, 2014. doi: 10.1016/j.ygyno.2013.12.014.

Gawande, Atul. "Two Hundred Years of Surgery." *New England Journal of Medicine* 366 (May 3, 2012): 1716–23. doi: 10.1056/NEJMra1202392.

Genetic Information Nondiscrimination Act of 2008, Pub. L. No. 110-233 (2008). www.gpo.gov/fdsys/pkg/PLAW-110publ233/html/PLAW-110publ233.htm.

Gessen, Masha. "A Medical Quest: In Which I Gain a Greater Understanding of Breasts and Ovaries." *Slate*, June 14, 2004.

Gessen, Masha. *Blood Matters: From Inherited Illness to Designer Babies, How the World and I Found Ourselves in the Future of the Gene.* New York: Mariner Books, Houghton Mifflin Harcourt, 2008.

Gitschier, Jane. "Evidence Is Evidence: An Interview with Mary-Claire King." *PLOS Genetics* 9, no. 9, September 26, 2013: e1003828. doi:10.1371/journal.pgen.1003828.

Gold, E. Richard, and Julia Carbone. "Myriad Genetics: In the Eye of the Policy Storm." Supplement, *Genetics in Medicine* 12, no. 4 April 2010: S39–70.

Gonzales-Angulo, A. M., F. Morales-Vasquez, and G. N. Hortobagyi. "Overview of Resistance to Systemic Therapy in Patients with Breast Cancer." *Advances in Experimental Medicine and Biology*, 608 (2007): 1–22.

Gøtzsche, P. C. "Relation Between Breast Cancer Mortality and Screening Effectiveness: Systematic Review of the Mammography Trials." *Danish Medical Bulletin*, 58(3), (March, 2011): A4246.

Graeser, M. K., C. Engel, K. Rhiem, D. Gadzicki, U. Bick, K. Kast, U. G. Froster, et al. "Contralateral Breast Cancer Risk in BRCA1 and BRCA2

Mutation Carriers." *Journal of Clinical Oncology* 27, no. 35 (December 10, 2009): 5887–92. doi: 10.1200/JCO.2008.19.9430.

Guarneri, V., and P. F. Conte. "The Curability of Breast Cancer and the Treatment of Advanced Disease." *European Journal of Nuclear Medicine and Molecular Imaging*, 31 Suppl. (June 2004): S149-61.

Gurdin, Michael, and Gene A. Carlin. "Complications of Breast Implantations." *Plastic and Reconstructive Surgery* 40, no. 6 (1967): 530–33.

Haiken, Elizabeth. *Venus Envy: A History of Cosmetic Surgery*. Baltimore: The Johns Hopkins University Press, 1997.

Hajdu, S. "A Note from History: Landmarks in History of Cancer, Part 1." *Cancer* 117 (October 19, 2010): 1097–1102. doi: 10.1002/cncr.25553.

Hall, J. M., M. K. Lee, B. Newman, J. E. Morrow, L. A. Anderson, B. Huey, and M. C. King. "Linkage of Early-Onset Familial Breast Cancer to Chromosome 17q21." *Science* 250, December 21, 1990: 1684–89.

Hartmann, L. C., T. A. Sellers, D. J. Schaid, T. S. Frank, C. L. Soderberg, D. L. Sitta, M. H. Frost, et al. "Efficacy of Bilateral Prophylactic Mastectomy in BRCA1 and BRCA2 Gene Mutation Carriers." *Journal of the National Cancer Institute* 93, no. 21 (November 7, 2001): 1633–37.

Hemlow, Joyce, ed. *Fanny Burney: Selected Letters and Journals*. Oxford: Clarendon Press, 1986.

Imber, Gerald. *Genius on the Edge: The Bizarre Double Life of Dr. William Stewart Halsted*. New York: Kaplan Publishing, 2010.

Ingham, S. L., M. Sperrin, and D. G. Evans. "Risk-Reducing Surgery Increases Survival in BRCA1/2 Mutation Carriers Unaffected at Time of Family Referral." *Breast Cancer Research and Treatment* 142, no. 3 (December 2013): 611–18.

Jacobsen, Oluf. *Heredity in Breast Cancer: A Genetic and Clinical Study of Two Hundred Probands*. English trans. Robert Fraser. Copenhagen: Nyt Nordisk Forlag, Arnold Busck, 1946.

Jacobson, Nora. *Cleavage: Technology, Controversy, and the Ironies of the Man-Made Breast*. New Brunswick, New Jersey: Rutgers University Press, 2000.

Jaffe, S. "Mary-Claire King." *Scientist* 18, no. 5, March 15, 2004. Accessed March 8, 2014. www.the-scientist.com/?articles.view/articleNo/15482/title/Mary-Claire-King.

Kasper, Anne S. and Susan J. Ferguson. *Breast Cancer: Society Shapes an Epidemic*. New York: St. Martin's Press, 2000.

King, Mary-Claire. "'The Race' to Clone BRCA1." *Science*, vol. 343 no. 6178, March 24, 2014: 1462–1465.

King, M., S. Wieand, K. Hale, M. Lee, T. Walsh, K. Owens, J. Tait, et al. "Tamoxifen and Breast Cancer Incidence Among Women with Inherited Mutations in BRCA1 and BRCA2: National Surgical Adjuvant Breast and Bowel Project (NSABP-P1) Breast Cancer Prevention Trial." *Journal of the American Medical Association* 286, no. 18 (November 14, 2001): 2251–56.

King, M. C., J. H. Marks, J. B. Mandell, and New York Breast Cancer Study Group. "Breast and Ovarian Cancer Risks Due to Inherited Mutations in BRCA1 and BRCA2." *Science* 302, no. 5645 (October 24, 2003): 643–46.

Kösters, J. P. and P. C. Gøtzsche. "Regular Self-Examination or Clinical Examination for Early Detection of Breast Cancer." *Cochrane Database of Systematic Reviews* 2 (2003): CD003373.

Lathan, S. Robert. "Dr. Halsted at Hopkins and at High Hampton." Baylor University Medical Center Proceedings 23, no. 1 (January 2010): 33–37.

Leopold, Ellen. *A Darker Ribbon: Breast Cancer, Women, and Their Doctors in the Twentieth Century*. Boston: Beacon Press, 1999.

Lord, S. J., W. Lei, P. Craft, J. N. Cawson, I. Morris, S. Walleser, A. Griffiths, et al. "A Systematic Review of the Effectiveness of Magnetic Resonance Imaging (MRI) as an Addition to Mammography and Ultrasound in Screening Young Women at High Risk of Breast Cancer." *European Journal of Cancer* 43, no. 13 (September 2007): 1905–17.

Loukas, M., R. S. Tubbs, N. Mirzayan, M. Shirak, A. Steinberg, and M. M. Shoja. "The History of Mastectomy." *American Surgeon* 77.5 (May 2011): 566–71.

Löwy, Ilana. *Preventive Strikes: Women, Precancer, and Prophylactic Surgery*. Baltimore: The Johns Hopkins University Press, 2010.

Metcalfe, K., S. Gershman, P. Ghadirian, H. T. Lynch, C. Snyder, N. Tung, C. Kim-Sing, et al. "Contralateral Mastectomy and Survival after Breast Cancer in Carriers of BRCA1 and BRCA2 Mutations: Retrospective Analysis." *BMJ* (February 11, 2014): 348:g226.

Moslehi, R., R. Singh, L. Lessner, and J. M. Friedman. "Impact of BRCA Mutations on Female Fertility and Offspring Sex Ratio." *American Journal of Human Biology* 22, no. 2 (March–April 2010): 201–5.

Mukherjee, Siddhartha. *The Emperor of All Maladies: A Biography of Cancer*. New York: Scribner, 2010.

National Library of Medicine. "The Edwin Smith Surgical Papyrus." Accessed March 6, 2014. http://archive.nlm.nih.gov/proj/ttp/flash/smith/smith.html.

Noonan, Kevin E. "Preliminary Injunction in Myriad v. Ambry and Gene-by-Gene: Myriad replies." Patent Docs, October 9, 2013. Accessed March 6, 2014. www.patentdocs.org/2013/10/preliminary-injunction-in-myriad-v-ambry-and-gene-by-gene-myriad-replies.html.

Oeters, Walter, and Victor Fornasier. "Complications from Injectable Materials Used for Breast Augmentation." *Canadian Journal of Plastic Surgery* 17, no. 3 (Fall 2009): 89–96.

Oken, Donald. "What to Tell Cancer Patients: A Study of Medical Attitudes." *Journal of the American Medical Association*. 175, no. 13 (April 1, 1961): 1120–28.

Oktay, Kutluk, J. Y. Kim, D. Barad, and S. N. Babayev. "Association of BRCA1 Mutations With Occult Primary Ovarian Insufficiency: A Possible Explanation for the Link Between Infertility and Breast/Ovarian Cancer Risks." *Journal of Clinical Oncology* 28, no. 2 (January 10, 2010): 240–44.

Olson, James S. *Bathsheba's Breast: Women, Cancer & History*. Baltimore: The Johns Hopkins University Press, 2002.

Pal, T., D. Keefe, P. Sun, S. A. Narod, and Hereditary Breast Cancer Clinical Study Group. "Fertility in Women with BRCA Mutations: A Case-Control Study." *Fertility and Sterility* 93, no. 6 (April 2010): 1805–8.

Passaperuma, K., E. Warner, and P. A. Causer. "Long-Term Results of Screening with Magnetic Resonance Imaging in Women with BRCA Mutations." *British Journal of Cancer* 107, no. 1 (June 26, 2012): 24–30.

Patenaude, Andrea Farkas. *Genetic Testing for Cancer: Psychological Approaches for Helping Patients and Families*. Washington, DC: American Psychological Association, 2005.

Penn, Jack. "A Perspective and Technique for Breast Reduction," in *Plastic and Reconstructive Surgery of the Breast*, edited by Robert M. Goldwyn. Boston: Little Brown, 1976.

Pennisi, Vincent R., and Angelo Capozzi. "The Incidence of Obscure Carcinoma in Subcutaneous Mastectomy." *Plastic and Reconstructive Surgery* 56, no. 1 (1975): 9–11.

Pijpe, A., N. Andrieu, D. F. Easton, A. Kesminiene, E. Cardis, C. Noguès, M. Gauthier-Villars, et al. "Exposure to Diagnostic Radiation and Risk of Breast Cancer Among Carriers of BRCA1/2 Mutations: Retrospective Cohort Study (GENE-RAD-RISK)." *BMJ* (September 6, 2012): 345:e5660.

Rebbeck, T. R., T. Friebel, H. T. Lynch, S. L. Neuhausen, L. van't Veer, J. E. Garber, G. R. Evans, et al. "Bilateral Prophylactic Mastectomy Reduces Breast Cancer Risk in BRCA1 and BRCA2 Mutation Carriers: The PROSE Study Group." *Journal of Clinical Oncology* 22, no. 6 (March 15, 2004): 1055–62.

Rocca, W. A., B. R. Grossardt, M. de Andrade, G. D. Malkasian, and L. J. Melton 3rd. "Survival Patterns after Oophorectomy in Premenopausal Women: A Population-Based Cohort Study." *Lancet Oncology* 7, no. 10 (October 2006): 821–28.

Rudnick, Joanna. *In the Family*. Film. Directed by Joanna Rudnick. Chicago: Kartemquin Films, 2008.

Sandison, A. T. "The First Recorded Case of Inflammatory Mastitis—Queen Atossa of Persia and the Physician Democêdes." *Medical History* 3, no. 4 (October 1959): 317–22.

Schwartz, M. D., C. Isaacs, K. D. Graves, B. N. Peshkin, C. Gell, C. Finch, S. Kelly, et al. "Long-Term Outcomes of BRCA1/BRCA2 Testing: Risk

Reduction and Surveillance." *Cancer* 118, no. 2 (January 15, 2012): 510–17. doi: 10.1002/cncr.26294.

Shuster, Lynne T., Bobbie S. Gostout, Brandon R. Grossardt, and Walter A. Rocca. "Prophylactic Oophorectomy in Pre-menopausal Women and Long-Term Health—A Review." *Menopause International* 14, no. 3 (2008): 111–16.

Simoni, Robert D., Robert L. Hill, and Martha Vaughn. "The Discovery of Estrone, Estriol, and Estradiol and the Biochemical Study of Reproduction. The Work of Edward Adelbert Doisy." *Journal of Biological Chemistry* 277, e17 (July 12, 2002).

Singh, K., J. Lester, B. Karlan, C. Bresee, T. Geva, and O. Gordon. "Impact of Family History on Choosing Risk-Reducing Surgery among BRCA Mutation Carriers." *American Journal of Obstetrics and Gynecology* 208, no. 4 (April 2013): 329.e1-6.

Skolnick, Mark. "Declaration of Mark Skolnick," AMP et al. v. Myriad Genetics. United States District Court for the Southern District of New York. December 23, 2009: No. 09 Civ. 4515.

Sleeman, Jonathan, and Patricia S. Steeg. "Cancer Metastasis as a Therapeutic Target." *European Journal of Cancer* 46, no. 7 (May 2012): 1177–80.

Stark, Lizzie. "Goodbye to My Breasts." *Daily Beast*, April 24, 2010. www.thedailybeast.com/articles/2010/04/24/good-bye-to-my-breasts.html.

Tavtigian, S. V., J. Simard, J. Rommens, F. Couch, D. Shattuck-Eidens, S. Neuhausen, S. Merajver, et al. "The Complete BRCA2 Gene and Mutations in 13q-Linked Kindreds." *Nature Genetics* 12, (March, 1996): 1–6.

Thomas, D. B., D. L. Gao, R. M. Ray, W. W. Wang, C. J. Allison, F.L. Chen, P. Porter, et al. "Randomized Trial of Breast Self-Examination in Shanghai: Final Results." *Journal of the National Cancer Institute* 94, no. 19 (October 2, 2002): 1445–57.

Thomas, T. G. "On the Removal of Benign Tumors of the Mamma without Mutilation of the Organ." *N.Y. State J. Med. Obstet. Rev.* (April 1882): 337–40.

van der Groep, Petra, Elsken van der Wall, and Paul J. van Diest. "Pathology of Hereditary Breast Cancer." *Cellular Oncology* (Dordrecht) 34, no. 2, April 2011: 71–88.

van Sprundel, T. C., M. K. Schmidt, M. A. Rookus, R. Brohet, C. J. van Asperen, E. J. Rutgers, L. J. Van't Veer, et al. "Risk Reduction of Contralateral Breast Cancer and Survival after Contralateral Prophylactic Mastectomy in BRCA1 or BRCA2 Mutation Carriers." *British Journal of Cancer* 93, no. 3 (2005): 287–92.

Waldholz, Michael. *Curing Cancer: The Story of the Men and Women Unlocking the Secrets of Our Deadliest Illness*. New York: Simon & Schuster, 1997.

Welch, Gilbert, and Brittney A. Frankel. "Likelihood That a Woman with Screen-Detected Breast Cancer Has Had Her 'Life Saved' by That Screening." *Archives of Internal Medicine* 171, no. 22 (December 11, 2011), 2043–46.

Wheelwright, Jeff. *The Wandering Gene and the Indian Princess: Race, Religion, and DNA*. New York: W.W. Norton & Company, 2012.

Williams, Florence. *Breasts: A Natural and Unnatural History*. New York: W.W. Norton & Company, 2012.

Wyklicky, Helmut, and Manfred Skopec. "Ignaz Philipp Semmelweis, the Prophet of Bacteriology." *Infection Control* 4, no. 5 (Sep.–Oct., 1983): 367–70.

Yalom, Marilyn. *A History of the Breast*. New York: Ballantine Books, 1997.

Notes

Prologue

Every person has two pairs of BRCA genes: Location of BRCA genes: "Hereditary Breast Ovarian Cancer Syndrome (BRCA1/BRCA2)," Stanford Medicine, Cancer Institute, accessed March 6, 2014, http://cancer.stanford.edu/information/geneticsAndCancer/types/herbocs.html.

Scientists identified these genes in the 1990s: William Check, "BRCA: What We Now Know," *CAP Today*, September 2006, www.cap.org/apps/cap.portaL?_nfpb=true&cntvwrPtlt_actionOverride=%2Fportlets%2FcontentViewer%2Fshow&_windowLabel=cntvwrPtlt&cntvwrPtlt%7BactionForm.contentReference%7D=cap_today%2Ffeature_stories%2F0906BRCA.html&_state=maximized&_pageLabel=cntvwr; Sarah Berger, "Genetics Pioneer Mary Claire King," *Moment*, accessed March 6, 2014. www.momentmag.com/2013-guide-to-jewish-genetic-diseases.

Lifetime chance of developing cancer chart: "BRCA1 and BRCA2: Cancer Risk and Genetic Testing," National Cancer Institute Fact Sheet, accessed March 6, 2014, www.cancer.gov/cancertopics/factsheet/Risk/BRCA; Similar numbers at "Hereditary Breast Ovarian Cancer Syndrome (BRCA1/BRCA2)."

In addition to having an elevated lifetime risk: Younger ages: National Cancer Institute "BRCA1 and BRCA2: Cancer Risk and Genetic Testing."

The median age of breast cancer diagnosis: "SEER Stat Fact Sheets: Breast Cancer," Surveillance, Epidemiology, and End Results Program, National Cancer Institute, National Institutes of Health, accessed March 6, 2014, http://seer.cancer.gov/statfacts/html/breast.html.

Triple-negative cancers are resistant: "BRCA1 and BRCA2: Cancer Risk and Genetic Testing."

BRCA mutations are pretty rare: "Genetics of Breast and Ovarian Cancer (PDQ®)," National Cancer Institute, National Institutes of Health, accessed March 6, 2014, www.cancer.gov/cancertopics/pdq/genetics/breast-and-ovarian/HealthProfessional/page2#Reference2.21; "Hereditary Breast Ovarian Cancer Syndrome (BRCA1/BRCA2)."

yet this tiny segment accounts for fully 5 to 10 percent: "BRCA1 and BRCA2: Cancer Risk and Genetic Testing."

The BRCA1 and BRCA2 genes produce proteins: Tumor-suppressor genes: Stanford Medicine, "Hereditary Breast Ovarian Cancer Syndrome (BRCA1/BRCA2)"; Repairing cellular damage: National Cancer Sheet "BRCA1 and BRCA2: Cancer Risk and Genetic Testing."

Since as we age we all acquire genetic mutations: For an explanation of genetics and how cancer grows, see: "What Is Cancer?," American Cancer Society, accessed March 6, 2014, www.cancer.org/cancer/cancerbasics/what-is-cancer; For an explanation of how tumor suppressors cause cancer, see: "How Genes Cause Cancer," Stanford Medicine, Cancer Institute, accessed March 6, 2014, http://cancer.stanford.edu/information/geneticsAndCancer/genesCause.html.

Typically, the women who choose surgery: C. Garcia, et. al, "Risk Management Options Elected by Women after Testing Positive for a BRCA Mutation," *Gynecologic Oncology* 132, no. 2 (February 2014): 428–33, accessed March 6, 2014, doi: 10.1016/j.ygyno.2013.12.014; Different numbers that follow the same general trend can be found here: M. S. Beattie, et al., "Uptake, Time Course, and Predictors of Risk-Reducing Surgeries in BRCA Carriers," *Genetic Testing and Molecular Biomarkers* 13, no. 1 (February 2009): 51–56, doi: 10.1089/gtmb/2008.0067; K. Singh, et al., "Impact of Family History on Choosing Risk-Reducing Surgery among BRCA Mutation Carriers," *American Journal of Obstetrics and Gynecology* 208, no. 4 (April 2013): 329.e1-6, doi: 10.1016/j.ajog.2013.01.026; M. D. Schwartz, et al., "Long-Term Outcomes of BRCA1/BRCA2 Testing: Risk Reduction and Surveillance," *Cancer* 118, no. 2 (January 15, 2012): 510–17, doi: 10.1002/cncr.26294.

Already, we know of other genetic mutations: "Genetic Testing for Hereditary Cancer Syndromes," National Cancer Institute, National Institutes of Health, accessed March 6, 2014, www.cancer.gov/cancertopics/factsheet/Risk/genetic-testing; Certain types of arthritis: "Heredity and Arthritis," American College of Rheumatology, accessed March 6, 2014, www.rheumatology.org/Practice/Clinical/Patients/Diseases_And_Conditions/Heredity_and_Arthritis; Diabetes: "Type 1 Diabetes," Genetics Home Reference, National Library of Medicine, accessed March 6, 2014, http://ghr.nlm.nih.gov/condition/type-1-diabetes; Asthma: "Genes and Disease: Asthma," National Center for Biotechnology Information, National Institutes for Health, accessed March 6, 2014, www.ncbi.nlm.nih.gov/books/NBK22181; Alzheimer's: "Alzheimer's Disease Genetics Fact Sheet," National Institute on Aging, National

Institutes of Health, accessed March 6, 2014, www.nia.nih.gov/alzheimers/publication/alzheimers-disease-genetics-fact-sheet.

Chapter 1: The Ham Speaks for Itself

The cancer is assigned a stage: "Breast Cancer Survival Rates by Stage," American Cancer Society, accessed March 6, 2014, www.cancer.org/cancer/breastcancer/detailedguide/breast-cancer-survival-by-stage.

one in two American men and one in three American women: "Lifetime Risk of Developing or Dying from Cancer," American Cancer Society, accessed March 6, 2014, www.cancer.org/cancer/cancerbasics/lifetime-probability-of-developing-or-dying-from-cancer. Additionally, one in four men and one in five women will die from cancer.

Chapter 2: "It's Everywhere"

Anthropologist Louis Leakey: Siddhartha Mukherjee, *The Emperor of All Maladies: A Biography of Cancer* (New York: Scribner, 2010), 42–45.

In his book The Emperor of All Maladies*:* Mukherjee, *Emperor of All Maladies*, 39–41; "The Edwin Smith Surgical Papyrus," National Library of Medicine, National Institutes of Health, accessed March 6, 2014, http://archive.nlm.nih.gov/proj/ttp/flash/smith/smith.html.

Hippocrates named this inexorable illness karkinos*:* Ira Flatow, "Science Diction: The Origin of the Word 'Cancer,'" *National Public Radio*, October 22, 2010, accessed March 6, 2014, www.npr.org/templates/story/story.php?storyId=130754101; S. Hajdu, "A Note from History: Landmarks in History of Cancer, Part 1," *Cancer* 117 (October 19, 2010): 1097–1102, doi: 10.1002/cncr.25553.

historical records have no shortage: The details about Atossa come from these accounts: James S. Olson, *Bathsheba's Breast: Women, Cancer and History* (Baltimore: Johns Hopkins University Press, 2002), 1–4; Mukherjee, *Emperor of All Maladies*, 42; A. T. Sandison, "The First Recorded Case of Inflammatory Mastitis—Queen Atossa of Persia and the Physician Democêdes," *Medical History* 3, no. 4 (October 1959): 317–22. The account of Anne of Austria comes from Olson, *Bathsheba's Breast*, 25.

Before he became a monster, Adolf Hitler: Olson, *Bathsheba's Breast*, 94–95.

the Lincoln cure: "Medicine: Sequel," *Time* 59, no. 20 (May 19, 1952): 77. Accessed March 3, 2014.

the concept of humors: Olson, *Bathsheba's Breast*, 12; Mukherjee, *Emperor of All Maladies*, 48.

Galenic theory treated illness: Mukherjee, *Emperor of All Maladies*, 48–49.

Due to Galen's influence: Hajdu, "A Note from History."

The anatomists who mapped the human body: Mukherjee, *Emperor of All Maladies*, 51–54.

In an era before doctors understood germs: Gerald Imber, *Genius on the Edge: The Bizarre Double Life of Dr. William Stewart Halsted* (New York: Kaplan Publishing, 2010), 114; Atul Gawande, "Slow Ideas: Some Innovations Spread Fast. How Do You Speed the Ones That Don't?," *New Yorker,* July 29, 2013, www.newyorker.com/reporting/2013/07/29/130729fa_fact_gawande.

antisepsis was born: Olson, *Bathsheba's Breast,* 48–49; Helmut Wyklicky and Manfred Skopec, "Ignaz Philipp Semmelweis, the Prophet of Bacteriology," *Infection Control* 4, no. 5 (Sep.–Oct., 1983): 367–70.

After months of dosing cats and dogs: Olson, *Bathsheba's Breast,* 53; Atul Gawande, "Two Hundred Years of Surgery," *New England Journal of Medicine* 366 (May 3, 2012): 1716–23, doi: 10.1056/NEJMra1202392.

advances in cellular theory: Mukherjee, *Emperor of All Maladies,* 13–16; Olson, *Bathsheba's Breast,* 57–58.

Enter the epic surgeon William Halsted: For information on Halsted, see Mukherjee, *Emperor of All Maladies,* 60–72; Olson, *Bathsheba's Breast,* 58–64. For a detailed account of Halsted's life, see: Imber, *Genius on the Edge.*

Born in 1852 to a wealthy family: Imber, *Genius on the Edge,* 250; Mukherjee, *Emperor of All Maladies,* 62–64; S. Robert Lathan, "Dr. Halsted at Hopkins and at High Hampton," *Baylor University Medical Center Proceedings* 23, no. 1 (January 2010): 33–37, accessed March 10, 2014, www.ncbi.nlm.nih.gov/pmc/articles/PMC2804495/.

English surgeon Charles Moore: Olson, *Bathsheba's Breast,* 60; Mukherjee, *Emperor of All Maladies,* 64.

Halsted advocated an operation: Mukherjee, *Emperor of All Maladies,* 64–65; Olson, *Bathsheba's Breast,* 60–61.

Halsted presented a stunning paper: Olson, *Bathsheba's Breast,* 62–63.

a "macabre marathon": Mukherjee, *Emperor of All Maladies,* 65.

"but for her, a far less aggressive procedure . . .": Mukherjee, *Emperor of All Maladies,* 67.

lumpectomy plus radiation was just as effective: Mukherjee, *Emperor of All Maladies,* 75–79, 195–201; Olson, *Bathsheba's Breast,* 86–91.

Early radiation therapy: Olson, *Bathsheba's Breast,* 90–91.

The carnage of war: Mukherjee, *Emperor of All Maladies,* 88; Olson, *Bathsheba's Breast,* 96.

He sent his results to Yale University: Mukherjee, *Emperor of All Maladies,* 89–90; Olson, *Bathsheba's Breast,* 96–98.

The mortality rate for ovarian cancer: "SEER Stat Fact Sheets: Ovary Cancer," Surveillance, Epidemiology, and End Results Program, National Cancer Institute, National Institutes of Health, accessed March 10, 2014, http://seer.cancer.gov/statfacts/html/ovary.html.

Scottish surgeon George Beatson: Olson, *Bathsheba's Breast*, 78–79; David D. Moore, "A Conversation with Elwood Jensen," *Annual Review of Physiology* 74 (March 2012): 1–11, accessed March 10, 2014, doi: 10.1146/annurev-physiol-020911-153327.

Canadian surgeon Charles Huggins: Robert D. Simoni, et al., "The Discovery of Estrone, Estriol, and Estradiol and the Biochemical Study of Reproduction. The Work of Edward Adelbert Doisy," *Journal of Biological Chemistry* 277, e17 (July 12, 2002), accessed March 10, 2014, www.jbc.org/content/277/28/e17.full; "Charles B. Huggins, MD, 1901–1997," University of Chicago Medicine, January 13, 1997, accessed March 10, 2014, www.uchospitals.edu/news/1997/19970113-huggins.html.

some breast cancers contain receptors: "Charles B. Huggins, MD, 1901–1997."

For these reasons, the National Comprehensive Cancer Network: Nicole Fawcett, "Most Women Who Have Double Mastectomy Don't Need It," Comprehensive Cancer Center, University of Michigan Health System, November 27, 2012, accessed March 7, 2014, www.cancer.med.umich.edu/news/unnecessary-double-mastectomy-2012.shtml; "Penn Researchers Find Contralateral Prophylactic Mastectomy (CPM) Offers Limited Gains to Life Expectancy for Breast Cancer Patients," Penn Medicine, University of Pennsylvania, December 7, 2011, accessed March 7, 2014, www.uphs.upenn.edu/news/News_Releases/2011/12/cpm-breast-cancer/; "Why Are Rates of Bilateral Mastectomies Rising?," Susan G. Komen for the Cure, accessed March 7, 2014, http://ww5.komen.org/Content.aspx?id=6442452097; Todd M. Tuttle, "Contralateral Prophylactic Mastectomy May Not Significantly Increase Life Expectancy in Women with Early-Stage Breast Cancer," Clinical Congress 2013, American College of Surgeons, October 7, 2013, accessed March 7, 2014, www.facs.org/clincon2013/press/tuttle.html; M. K. Graeser, et al., "Contralateral Breast Cancer Risk in BRCA1 and BRCA2 Mutation Carriers," *Journal of Clinical Oncology* 27, no. 35 (December 10, 2009): 5887–92, doi: 10.1200/JCO.2008.19.9430; William J. Gradishar, et al. "Breast Cancer." NCCN Clinical Practice Guidelines in Oncology (NCCN Guidelines ®), version 2.2014. Accessed March 7, 2014. www.nccn.org/professionals/physician_gls/pdf/breast.pdf.

Chapter 3: Gene Hunters

the very first Susan G. Komen Race: "Susan G. Komen Race for the Cure," Susan G. Komen, accessed March 11, 2014, http://ww5.komen.org/findarace.aspx.

Mary-Claire King is the Eleanor Roosevelt: Kevin Davies and Michael White, *Breakthrough: The Race to Find the Breast Cancer Gene*, (New York: John Wiley & Sons, 1995), 71.

Mary-Claire King puts a premium: " 'The Biggest Obstacle Is How to Have Enough Hours in the Day': Mary-Claire King," *Chronicle of Higher Education*, June 10, 2005, accessed March 8, 2014, http://chronicle.com/article/The-Biggest-Obstacle-Is-How/20248; Davies and White, *Breakthrough*, 72.

Mary-Claire King is fostering: "Israeli-Palestinian Genetic Research Project Advanced by US Donation," Fogarty International Center, National Institutes of Health, February 2010, accessed March 8, 2014, www.fic.nih.gov/news/globalhealthmatters/Pages/0210_genetics-israel-palestine.aspx; "2004 Genetics Prize: Mary-Claire King," Gruber Foundation, Yale University, accessed March 8, 2014, http://gruber.yale.edu/genetics/2004/mary-claire-king.

Wilson's theoretical and experimental discovery: Rebecca L. Cann, et al., "Mitochondrial DNA and Human Evolution," *Nature* (January 1, 1987): 31–36, doi: 10.1038/325031a0.

King and Wilson published their results: M. King and A. Wilson, "Evolution at Two Levels in Humans and Chimpanzees," *Science* 188, no. 4184 (April 1975): 107–116.

Working with Wilson taught her: Jane Gitschier, "Evidence Is Evidence: An Interview with Mary-Claire King," *PLOS Genetics* 9, no. 9 (September 26, 2013): e1003828, doi:10.1371/journal.pgen.1003828.

King was also deeply politically engaged: S. Jaffe, "Mary-Claire King," *Scientist* 18, no. 5 (March 15, 2004), accessed March 8, 2014, www.the-scientist.com/?articles.view/articleNo/15482/title/Mary-Claire-King.

Ralph Nader: Michael Waldholz, *Curing Cancer: The Story of the Men and Women Unlocking the Secrets of Our Deadliest Illness* (New York: Simon and Schuster, 1997), 97.

"I've learned not to question the motives of bastards": Thomas A. Bass, "The Gene Detective," *Observer*, October 17, 1993, D31.

King moved to Chile: "Putting the Puzzle Together," University of Washington, September 1996, accessed March 11, 2014, www.washington.edu/alumni/columns/sept96/king2.html; Smadar Reisfeld, "In Her DNA: Finding Lost Children, Linking Humans to Apes and Understanding Breast Cancer," *Haaretz*, January 17, 2013, accessed March 11, 2014; Waldholz, *Curing Cancer*, 100.

During the Dirty War of the 1970s: "Operation Condor: Cable Suggests U.S. Role," National Security Archive, George Washington University, March 6, 2001, accessed March 11, 2014, http://www2.gwu.edu/~nsarchiv/news/20010306/; John Dinges, *The Condor Years: How Pinochet and His Allies Brought Terrorism to Three Continents* (New York: New Press, 2004), 1; Marguerite Feitlowitz, *A Lexicon of Terror: Argentina and the Legacies of Torture, Revised and Updated with a New Epilogue*, (New York: Oxford University Press, 2011), 59, 66, 75, 78–79.

the Abuelas de Plaza de Mayo: "History of Abuelas de Plaza de Mayo," Abuelas de Plaza de Mayo, accessed March 11, 2014, www.abuelas.org.ar /english/history.htm; Davies and White, *Breakthrough*, 61.

blood proteins: "Using Genetics for Human rights," *Daily of the University of Washington*, May 12, 1997, accessed March 11, 2014, http://dailyuw.com /archive/1997/05/12/imported/using-genetics-human-rights#.UidSR WTwIad. Mitochondrial DNA: Mary-Claire King, "Statement of Mary-Claire King of the Human Genome Diversity Project to the National Academy of Sciences, 1996," Morrison Institute for Population and Resource Studies, Stanford University, September 16, 1996, accessed March 11, 2014, http://hsblogs.stanford.edu/morrison/2011/03/10 /statement-of-mary-clare-king-of-the-human-genome-diversity-project -to-the-national-academy-of-sciences-1996; "Encontramos a la nieta 110," Abuelas de Plaza de Mayo, February 6, 2014, accessed March 11, 2014, http:// www.abuelas.org.ar/comunicados/restituciones/res140206_1040-1.htm.

The identity of many of these children: Natalie Angier, "Scientist at Work: Mary-Claire King; Quest for Genes and Lost Children," *New York Times*, April 27, 1993, accessed March 11, 2014, www.nytimes.com /1993/04/27/science/scientist-at-work-mary-claire-king-quest-for-genes -and-lost-children.html?pagewanted=4&src=pm

Watch out, murderous dictator bastards: Goethe-Universität, "Geneticist Mary-Claire King Receives the 2013 Paul Ehrlich Prize" (press release), March 14, 2013, www.eurekalert.org/pub_releases/2013-03/guf-gmk 031413.php; King, speech to National Academy of Sciences, Morrison Institute, 1996.

The link between certain families and cancer: Petra van der Groep, et al., "Pathology of Hereditary Breast Cancer," *Cellular Oncology (Dordrecht)* 34, no. 2 (April 2011): 71–88, accessed March 13, 2014, www.ncbi.nlm .nih.gov/pmc/articles/PMC3063560/#CR24; Waldholz, *Curing Cancer*, 105.

These numbers convinced Broca: Anne J. Krush, "Contributions of Pierre Paul Broca to Cancer Genetics," Paper 316, *Transactions of the Nebraska Academy of Sciences and Affiliated Societies* (1979), http://digital commons.unl.edu/cgi/viewcontent.cgi?article=1315&context=tnas.

An early study on hereditary breast cancer: Oluf Jacobsen, *Heredity in Breast Cancer: A Genetic and Clinical Study of Two Hundred Probands*, English trans. Robert Fraser (Copenhagen: Nyt Nordisk Forlag, Arnold Busck, 1946).

The study showed that women with a first-degree relative: Ilana Löwy, *Preventive Strikes: Women, Precancer, and Prophylactic Surgery* (Baltimore: Johns Hopkins University Press, 2010), 170.

Today, about one in three women: American Cancer Society, "Lifetime Risk of Developing or Dying from Cancer."

"Researchers initially confused two distinct phenomena": Löwy, *Preventive Strikes*, 170.

Omaha physician Henry Lynch: Löwy, *Preventive Strikes*, 173; Waldholz, *Curing Cancer*, 107; Masha Gessen, *Blood Matters: From Inherited Illness to Designer Babies, How the World and I Found Ourselves in the Future of the Gene* (New York: Mariner Books, 2008), 119; Emily Wax-Thibodeaux, "As Helen Hunt Plays Her in a Movie, the Real Mary-Claire King Still Studies Breast Cancer," *Washington Post*, October 28, 2013.

geneticists had already figured out: Löwy, *Preventive Strikes*, 172.

She had a personal connection: Waldholz, *Curing Cancer*, 95; Bass, "Gene Detective."

"I was interested in it . . .": Emily Stone, "Mary-Claire King: Molecular Explorer," *Chicago Tribune*, June 25, 2004.

King speculates that the public feels: Ibid.

King had a eureka moment: Waldhoz, *Curing Cancer*, 101–102.

"After I had accepted the job. . .": Gitschier, "Evidence is Evidence."

"In retrospect, there was something liberating . . .": Wax-Thibodeaux, "Helen Hunt."

"It wasn't too long . . .": Gitschier, "Evidence Is Evidence."

Only a few hundred genes: Waldholz, *Curing Cancer*, 103. Number of human genes: David Bodine, "2012 National DNA Day Online Chatroom Transcript," National Human Genome Research Institute, National Institutes of Health, accessed March 10, 2014, www.genome.gov/DNADay/q.cfm?aid=2&year=2012; Douglas Main, "Humans May Have Fewer Genes Than Worms," *Popular Science*, January 3, 2014, accessed March 10, 2014.

she needed a bunch of huge cancer families: Waldholz, *Curing Cancer*, 107.

King's interview was rebroadcast: Gitschier, "Evidence Is Evidence."

King used some of the sequencing: Waldholz, *Curing Cancer*, 107.

If her lab could identify a good marker: Waldholz, *Curing Cancer*, 107–109.

Eventually, she and her team: J. M. Hall, et al., "Linkage of Early-Onset Familial Breast Cancer to Chromosome 17q21," *Science* 250 (December 21, 1990): 1684–89.

"of course everything was done by hand . . .": Gitschier, "Evidence Is Evidence."

And as many as one in two hundred women: Waldholz, *Curing Cancer*, 46–47.

"I presented our data . . .": Gitschier, "Evidence Is Evidence."

"He presented what I interpreted . . .": Ibid.

Lenoir had confirmed her results: Waldholz, *Curing Cancer*, 55.

King's news fired off: Ibid., 93.

Later, in her 2014 piece: Mary-Claire King. " 'The Race' to Clone BRCA1," *Science*, vol. 343 no. 6178, March 24, 2014: 1462–1465.
King's results were astonishing: Ibid., 46; Davies and White, *Breakthrough*, 132.
"If you consider human DNA . . .": Waldholz, *Curing Cancer*, 60–61.
the Human Genome Project had just begun: King, " 'The Race' to Clone BRCA1," 1462.
Mark Skolnick's lab in Utah: Ibid., 115; Davies and White, *Breakthrough*, 182–83.
The Mormon family trees: Tim B. Heaton, "Vital Statistics," *Encyclopedia of Mormonism*, Harold B. Lee Library, Brigham Young University, accessed March 13, 2014, http://eom.byu.edu/index.php/Vital_Statistics.
"Discovering that you're related . . .": "Family History," Church of Jesus Christ of Latter-Day Saints, accessed March 13, 2014, http://mormon.org/values/family-history.
tracing family trees is a holy mission: Ibid.
The potential juxtaposition: Davies and White, *Breakthrough*, 185–86.
He decided to investigate: Ibid., 186; Waldholz, *Curing Cancer*, 118–19.
his team spent the next five years: Davies and White, *Breakthrough*, 187.
As Skolnick identified possible cancer families: Ibid.
Skolnick went after a simpler hereditary disease: Ibid., 188.
He focused on hemochromatosis: Waldholz, *Curing Cancer*, 122–23.
Researchers already knew it was hereditary: Davies and White, *Breakthrough*, 189.
Building on French research: Ibid., 190.
The discovery of RFLPs: Waldholz, *Curing Cancer* 127–30.
Mark Skolnick joined forces: Mark Skolnick, "Why He Formed Myriad," video, DNA Learning Center, accessed March 13, 2014, www.dnalc.org/view/15244-Why-he-formed-Myriad-Mark-Skolnick.html.
"I was also keenly aware . . .": "Declaration of Mark Skolnick," AMP et al. v. Myriad Genetics, United States District Court for the Southern District of New York, No. 09 Civ. 4515, December 23, 2009, 5; Waldholz, *Curing Cancer*, 226.
If Skolnick couldn't get the money: "Declaration of Mark Skolnick," 6; Waldholz, *Curing Cancer*, 226.
"We were acutely aware . . .": "Declaration of Mark Skolnick."
"People ask me how I felt": Waldholz, *Curing Cancer*, 231.
A 1994 New York Times article: Natalie Angier, "Fierce Competition Marked Fervid Race for Cancer Gene," *New York Times*, September 20, 1994, C1.
Later, she would tell the Lancet*:* Geoff Watts, "Mary-Claire King: Taking Genes beyond the Lab," *Lancet* 382, no. 9887 (July 2013): 119, www.thelancet.com/journals/lancet/article/PIIS0140-6736(13)61551-2/fulltext.

With its commercial motives, Myriad: Associated Press, "Dispute Arises over Patent for a Gene: U.S. Seeking Role in Cancer Therapies," *New York Times*, October 30, 1994, 32.

Though the patent disputes over BRCA: E. Richard Gold and Julia Carbone, "Myriad Genetics: In the Eye of the Policy Storm," supplement, *Genetics in Medicine* 12, no. 4 (April 2010): S39–70.

Stratton's group published their results: Gina Kolata, "Scientists Speedily Locate a Gene That Causes Breast Cancer; Better Screening Is Seen," *New York Times*, December 21, 1995, B18.

The day before Stratton's group's paper: Associated Press, "Second Breast Cancer Gene Identified," *Washington Post*, December 21, 1995, A24; "Myriad Files Patent for a Second Gene Causing Breast Cancer," *Wall Street Journal*, December 21, 1995, B8.

The company would publish: S. V. Tavtigian, et al., "The BRCA2 Gene and Mutations in 13q-Linked Kindreds," *Nature Genetics* 12 (1996): 1–6.

there is still plenty to learn: "Genetic Testing," Facing Our Risk of Cancer Empowered (FORCE), accessed March 10, 2014, www.facingourrisk.org/info_research/hereditary-cancer/genetic-testing/index.php?PHPSESSID=6ec891ded9cddf8cfcb757d419758706; "Inherited Risk for Breast and Ovarian Cancers," Memorial Sloan Kettering Cancer Center, accessed March 10, 2014, www.mskcc.org/cancer-care/hereditary-genetics/inherited-risk-breast-ovarian.

Chapter 4: Myriad's Monopoly

pay Myriad Genetics: Test cost: David B. Agus, "The Outrageous Cost of a Gene Test," *New York Times*, May 20, 2013, A25; Andre Pollack, "A Genetic Test That Very Few Need, Marketed to the Masses," *New York Times*, September 11, 2007.

The courts have interpreted patent law: "General Information Concerning Patents," United States Patent and Trademark Office, Department of Commerce, November 2011, accessed March 13, 2014, www.uspto.gov/patents/resources/general_info_concerning_patents.jsp.

inventors earn the right: Ibid.

help identify remains from the World Trade Center: Andrew Pollack, "Business; Identifying the Dead, 2,000 Miles Away," *New York Times*, September 30, 2001.

"There's no controversial patent . . .": Joanna Rudnick, *In the Family*, film, directed by Joanna Rudnick (Chicago: Kartemquin Films, 2008).

They noted that Myriad sent cease-and-desist: Ass'n for Molecular Pathology v. U.S. Patent and Trademark Office, No. 09 Civ. 4515, slip op. at 59–63 (S.D.N.Y. March 29, 2010).

Attorneys for Myriad would cite: Molecular Pathology, No. 09 Civ. 4515, slip op. at 76–77.

Myriad didn't permit researchers: Molecular Pathology, No. 09 Civ. 4515, slip op. at 72.

Myriad's foes also argued: Molecular Pathology, No. 09 Civ. 4515, slip op. at 65–66.

Myriad argued that it permitted some labs: Molecular Pathology, No. 09 Civ. 4515, slip op. at 67–68.

850 VUSs from twenty-one thousand patients: Molecular Pathology, No. 09 Civ. 4515, slip op. at 65–66.

Similarly, the plaintiffs argued that Myriad's monopoly: Molecular Pathology, No. 09 Civ. 4515, slip op. at 56–59.

free testing to uninsured patients: "Responsibility," Myriad Genetics, accessed March 13, 2014, www.myriad.com/responsibility/.

A 2014 opinion by a judge in another legal matter: Robert J. Shelby, "Memorandum Decision and Order Denying Plaintiff's Motion for Preliminary Injunction." University of Utah Research Foundation, et al. vs. Ambry Genetics Corporation. 2:13-CV-00640-RJS. U.S. District Court, Utah, Central Division. March 10, 2014, 2; Myriad Genetics Inc., "Written Comments on Genetic Diagnostic Testing Study Before United States Patent and Trademark Office (USPTO)," March 26, 2012, 21.

In the time it takes for a patent case to settle: Daniel J. Kevles, "Can They Patent Your Genes?," *New York Review of Books*, March 7, 2013. The suit was first filed in 2009.

Myriad's brief argued that "the established rule . . .": Gregory A. Castanias, "Brief for Respondents," AMP et al., v. Myriad Genetics et al., No. 12-398, March 7, 2013, 32. Parke-Davis & Co. v. H. K. Mulford Co.: Parke-Davis & Co. v. H. K. Mulford Co., 189 F. 95 (C.C.S.D.N.Y. 1911).

If you extract the molecule: Castanias, "Brief for Respondents," 23–29, 35–45.

Similarly, the brief argues, "Without Myriad's work . . .": Ibid., 35.

They also cited the American Fruit Growers*:* Ibid., 29.

So too, they argued, isolating a segment: Ibid., 30.

They cited the Gen. Elec. Co. v. De Forest Radio Co.*:* Ibid., 38–39.

"Myriad did not invent . . .": Ibid., 39.

The Justice Department argued that cDNA: Ibid., 19.

By early 2014 half a dozen companies: Kevin E. Noonan, "Preliminary Injunction in Myriad v. Ambry and Gene-by-Gene: Myriad replies," Patent Docs, October 9, 2013, www.patentdocs.org/2013/10/preliminary-injunction-in-myriad-v-ambry-and-gene-by-gene-myriad-replies.html; Personal interview with Robert Cook-Deegan.

In March 2014, a judge shot down: Robert J. Shelby, "Memorandum Decision and Order Denying Plaintiff's Motion for Preliminary Injunction."

Chapter 5: Positive

The project, begun in 1990: "Human Genome Project," National Institutes of Health, accessed March 7, 2014, www.report.nih.gov/NIHfactsheets/ViewFactSheet.aspx?csid=45&key=H#H; "Human Genome Project," Human Genome Project Information Archive 1990–2003, accessed March 13, 2014, www.ornl.gov/sci/techresources/Human_Genome/home.shtml.

"At one time, such medical clairvoyance . . .": "Health Insurance in the Age of Genetics," National Human Genome Research Institute, National Institutes of Health, July 1997, accessed March 13, 2014, www.genome.gov/10000879.

"We can now only barely imagine": William Jefferson Clinton, "President Clinton's Comments on the Signing of Executive Order 13145," National Human Genome Research Institute, National Institutes of Health, February 8, 2000, accessed March 13, 2014, www.genome.gov/10002346.

The head of the National Human Genome Research Institute: Francis Collins, "Director of NHGRI Applauds President Clinton's Action to Protect Federal Workers from Genetic Discrimination," National Human Genome Research Institute, National Institutes of Health, February 2000, accessed March 13, 2014, www.genome.gov/10002345.

As early as 1993, an NIH report: "Genetic Information and Health Insurance Report of the Task Force on Genetic Information and Insurance," NIH-DOE Working Group on Ethical, Legal and Social Implications of Human Genome Research, May 10, 1993, www.genome.gov/10001750.

The Department of Labor report: "Genetic Information and the Workplace," Department of Labor, Department of Health and Human Services, Equal Employment Opportunity Commission, Department of Justice, January 20, 1998, www.dol.gov/dol/aboutdol/history/herman/reports/genetics.htm.

In 1998, some employees of Lawrence Berkeley Laboratory: Norman Bloodsaw v. Lawrence Berkeley Laboratory, No. 96-16526, Court of Appeals, Ninth Circuit, June 10, 1997.

The US Equal Employment Opportunity Commission: Andrew Plemmons Pratt, "Data Bank: Consumer Genetic Testing and Cases of Genetic Discrimination," *Science Progress*, March 2, 2009, accessed March 13, 2014; http://scienceprogress.org/2009/03/data-bank-gina/; "EEOC Settles ADA Suit Against BNSF for Genetic Bias," press release, U.S. Equal Employment Opportunity Commission, April 18, 2001, accessed March 13, 2014, www.eeoc.gov/eeoc/newsroom/release/4-18-01.cfm.

The lawsuit settled: "EEOC Settles ADA Suit."

Legislators acted because the public feared: Francis Collins, "Statement of Francis Collins on President Clinton's Announcement to End Genetic

Discrimination in Health Insurance," National Human Genome Research Institute, National Institutes of Health, July 1997, accessed March 13, 2014, www.genome.gov/10000882. A study in 2000 found that 68 percent of genetic counselors themselves "would not bill their insurance companies for genetic testing because of fear of discrimination." E. T. Matloff, et al., "What Would You Do? Specialists' Perspectives on Cancer Genetic Testing, Prophylactic Surgery, and Insurance Discrimination," *Journal of Clinical Oncology* 18, no. 12 (June 2000): 2482–92; Committee on Health, Education, and Pensions, *Report on Genetic Information Nondiscrimination Act of 2007.* 110th Congress S. Rep. 110-48, 110th Cong. , 1st Sess., at 1–2, 5–7 (April 10. 2007).

In 1971, President Nixon dedicated $10 million: Richard Nixon, "Statement on Signing the National Sickle Cell Anemia Control Act," May 16, 1972, American Presidency Project, www.presidency.ucsb.edu/ws/?pid=3413.

But in the wake of the funding increase: K. G. Fulda and K. Lykens, "Ethical Issues in Predictive Genetic Testing: A Public Health Perspective," *Journal of Medical Ethics* 32, no. 3 (March 2006): 143–47.

Legislators recognized that genetic discrimination: Committee on Health, Education, and Pensions, *Report on Genetic Information Nondiscrimination Act of 2007.*

As journalist Jeff Wheelwright writes: Jeff Wheelwright, *The Wandering Gene and the Indian Princess: Race, Religion, and DNA* (New York: W.W. Norton and Company, 2012), 152; Karen H. Rothenberg and Amy B. Rutkin, "Toward a Framework of Mutualism: The Jewish Community in Genetics Research," *Community Genetics*, 1 (1998): 148–153.

Federal laws against genetic discrimination: "Genetic Discrimination," Genetics and Public Policy Center, accessed April 18, 2014, www.dnapolicy.org/gina/gina.history.html.

The new law became effective in 2009: Genetic Information Nondiscrimination Act of 2008, Pub. L. No. 110-233 (2008). www.gpo.gov/fdsys/pkg/PLAW-110publ233/html/PLAW-110publ233.htm.

Still, GINA does not cover: David Shultz, "It's Legal for Some Insurers to Discriminate Based on Genes," National Public Radio, January 17, 2013, accessed April 16, 2014, www.npr.org/blogs/health/2013/01/17/169634045/some-types-of-insurance-can-discriminate-based-on-genes.

Over the course of four years: "Genetic Information Non-Discrimination Act Charges FY 2010-2013," U.S. Equal Opportunity Commission, accessed March 13, 2014, www.eeoc.gov/eeoc/statistics/enforcement/genetic.cfm; "U.S. and World Population Clock," United States Census Bureau, accessed March 13, 2014, www.census.gov/popclock/.

Chapter 6: Watchful Waiting

I'd even heard horror stories: As the chief medical officer for the America Cancer Society reminds me, this lack of knowledge on the part of medical professionals, while frustrating and possibly life-threatening for some women, may be understandable, as "if you are a forty-five-year-old physician, this whole BRCA thing came about after you finished your residency."

According to the National Comprehensive Cancer Network: Mary B. Daly, et al., "Genetic/Familial High-Risk Assessment: Breast and Ovarian," (NCCN Clinical Practice Guidelines in Oncology (NCCN Guidelines ®), version 1.2014), National Comprehensive Cancer Network, accessed March 7, 2014, HBOC-A. www.nccn.org/professionals/physician_gls/pdf/genetics_screening.pdf.

Currently scientists are studying screening: "Screening and Testing to Detect Cancer," National Cancer Institute, National Institutes of Health, accessed March 13, 2014, www.cancer.gov/cancertopics/screening.

You can think of cancer screening like a net: Adapted from Mukherjee, *Emperor of All Maladies,* 291–94.

Siddhartha Mukherjee explains the complications: Mukherjee, *Emperor of All Maladies,* 292–93.

It used to be the leading cause: "Cervical Cancer Statistics," Centers for Disease Control and Prevention, December 20, 2012, accessed March 13, 2014, www.cdc.gov/cancer/cervical/statistics/; "United States Cancer Statistics: 2006–2010 Top Ten Cancers," U.S. Cancer Statistics Working Group, *United Cancer Statistics: 1999-2010 Incidence and Mortality Web-based Report* (Atlanta: U.S. Department of Health and Human Services, Centers for Disease Control and Prevention and National Cancer Institute, 2013), accessed March 13, 2014, http://apps.nccd.cdc.gov/uscs/toptencancers.aspx.

"His wife Maria, in surely one . . .": Mukherjee, *Emperor of All Maladies,* 286–87.

Papanicolaou began reading slides: Mukherjee, *Emperor of All Maladies,* 287–89; Löwy, *Preventive Strikes,* 89.

The Pap smear meant doctors: Mukherjee, *Emperor of All Maladies,* 289–90.

As Ilana Löwy, a senior researcher: Löwy, *Preventive Strikes,* 96, 128–129, 140–141; Robert A. Aronowitz, *Unnatural History: Breast Cancer and American Society* (New York: Cambridge University Press, 2007) 219–20; "Cervical Dysplasia," Medline Plus, National Library of Medicine, National Institutes of Health, updated February 8, 2013, accessed March 13, 2014, www.nlm.nih.gov/medlineplus/ency/article/001491.htm.

The Pap smear's success: Aronowitz, *Unnatural History,* 221.

Nonetheless, the same logic: Löwy, *Preventive Strikes,* 143; Aronowitz, *Unnatural History,* 220–21.

Berlin surgeon Albert Salomon: Mukherjee, *Emperor of All Maladies,* 290.
The method didn't resurface: Aronowitz, *Unnatural History,* 223.
A 1963 study of eighty thousand: Mukherjee, *Emperor of All Maladies,* 297.
Mukherjee called the Edinburgh study: Mukherjee, *Emperor of All Maladies,* 298.
"This pattern—a clearly discernible benefit . . .": Mukherjee, *Emperor of All Maladies,* 300–1.
a paper in the Journal of the National Cancer Institute*:* "Mammography Screening Shows Limited Effect on Breast Cancer Mortality in Sweden," *Science Daily,* July 17, 2012, accessed March 13, 2014, www.sciencedaily.com/releases/2012/07/120717162629.htm.
These ambiguities around risk: Löwy, *Preventive Strikes,* 146.
The numbers show that of the BRCA women: C. Garcia, et al., "Risk Management Options Elected by Women after Testing Positive for a BRCA Mutation." Different numbers that follow the same general trend can be found here: M. S. Beattie, et al., "Uptake, Time Course, and Predictors of Risk-Reducing Surgeries in BRCA Carriers."; K. Singh et al., "Impact of Family History on Choosing Risk-Reducing Surgery among BRCA Mutation Carriers."; M. D. Schwartz, et al., "Long-Term Outcomes of BRCA1/BRCA2 Testing: Risk Reduction and Surveillance."
A study that examined the breasts: Löwy, *Preventive Strikes,* 153.
"An enhancement of radiologists' capacity . . .": Ibid., 153.
Malcolm Gladwell posed one: Malcolm Gladwell, "The Picture Problem," *New Yorker,* December 13, 2004; National Cancer Insitute, "SEER Stat Fact Sheets: Breast Cancer."
If screening works: Given that recurrence rates are different for different stages and types of breast cancer, the exact percentage of early-stage cancers that later metastasize is difficult to pin down. Here are a few estimates: **Up to 30 percent**. (Elia Ben-Ari, "A Conversation with Dr. Patricia Steeg on Redesigning Clinical Trials to Test Therapies That Could Prevent Cancer Metastasis," *NCI Cancer Bulletin* 9, no. 12 (June 12, 2012), accessed March 13, 2014. www.cancer.gov/ncicancerbulletin/061212/page4). **21 to 25 percent** for estrogen-receptor positive cancer with five or ten years of tamoxifen treatement. (National Cancer Institute, "Ten Years of Tamoxifen Reduces Breast Cancer Recurrences, Improves Survival," March 20, 2013, accessed April 17, 2014, www.cancer.gov/clinicaltrials/results/summary/2013/tamoxifen-10yrs0313). **About 30 percent** of early-stage patients end up with recurrence. (A.M. Gonzales-Angulo, et al., "Overview of Resistance to Systemic Therapy in Patients with Breast Cancer," *Advances in Experimental Medicine and Biology,* 608 (2007): 1–22). **25 to 45 percent** of breast cancer patients develop metastatic cancer. (V. Guarneri and P. F. Conte, "The Curability of Breast Cancer and the Treatment of Advanced Disease," *European*

Journal of Nuclear Medicine and Molecular Imaging, 31 Suppl. (June, 2004): S149-61).

A 2011 Danish meta-study: P. C. Gøtzsche, "Relation Between Breast Cancer Mortality and Screening Effectiveness: Systematic Review of the Mammography Trials," *Danish Medical Bulletin*, 58(3), (March, 2011), A4246.

"Despite substantial increases . . .": Archie Bleyer and Gilbert Welch, "Effect of Three Decades of Screening Mammography on Breast Cancer Incidence," *New England Journal of Medicine*, November 22, 2012, 367: 1998–2005. A study by Dartmouth researchers says it's overwhelmingly likely that mammography does not save lives: Gilbert Welch and Brittney A. Frankel, "Likelihood That a Woman with Screen-Detected Breast Cancer Has Had Her 'Life Saved' by That Screening," *Archives of Internal Medicine* 171, no. 22 (December 11, 2011), 2043–46.

Mammograms may or may not: Wendy A. Berg, et al., "Detection of Breast Cancer with Addition of Annual Screening Ultrasound or a Single Screening MRI to Mammography in Women with Elevated Breast Cancer Risk," *Journal of the American Medical Association* 307, no. 13 (April 4, 2012); "American Cancer Society Recommendations for Early Breast Cancer Detection in Women without Breast Symptoms," American Cancer Society, January 28, 2014, accessed March 14, 2014, www.cancer.org/cancer/breastcancer/moreinformation/breastcancerearlydetection/breast-cancer-early-detection-acs-recs.

Still, so far, adding MRIs to mammograms: S. J. Lord, et al., "A Systematic Review of the Effectiveness of Magnetic Resonance Imaging (MRI) as an Addition to Mammography and Ultrasound in Screening Young women at High Risk of Breast Cancer," *European Journal of Cancer* 43, no. 13 (September 2007): 1905–17; K. Passaperuma, et al., "Long-Term Results of Screening with Magnetic Resonance Imaging in Women with BRCA Mutations," *British Journal of Cancer* 107, no. 1 (June 26, 2012): 24–30.

A mammogram exposes women: "General Questions and Comments on Radiation Risk," American Cancer Society, August 9, 2013, accessed March 14, 2014, www.cancer.org/treatment/understandingyourdiagnosis/examsandtestdescriptions/imagingradiologytests/imaging-radiology-tests-rad-risk.

"Quite honestly, there's a lot . . .": "Exposure to X-rays Raises Risk of Breast Cancer in Young Women with BRCA Faults," Cancer Research UK, September 6, 2012, accessed March 14, 2014, www.cancerresearchuk.org/about-us/cancer-news/press-release/exposure-to-x-rays-raises-risk-of-breast-cancer-in-young-women-with-brca-faults; A. Pijpe, et al., "Exposure to Diagnostic Radiation and Risk of Breast Cancer among Carriers of BRCA1/2 Mutations: Retrospective Cohort Study (GENE-RAD-RISK)," *BMJ* (September 6, 2012): 345:e5660.

Two studies performed in Chinese and Russian: D. B. Thomas, et al., "Randomized Trial of Breast Self-Examination in Shanghai: Final Results," *Journal of the National Cancer Institute* 94, no. 19 (October 2, 2002): 1445–57; J. P. Kösters and P. C. Gøtzsche, "Regular Self-Examination or Clinical Examination for Early Detection of Breast Cancer," *Cochrane Database of Systematic Reviews* 2 (2003): CD003373.

Chapter 7: A Tale of Too Many Mastectomies

They're even named for their breastlessness: Marilyn Yalom, *A History of the Breast* (New York: Ballantine Books, 1997), 22–23.

Saint Agatha, a Sicilian virgin: Yalom, *History of the Breast*, 32–36.

In 1810, she developed a lump: Fanny Burney, "A Mastectomy," in *Fanny Burney: Selected Letters and Journals*, ed. Joyce Hemlow (Oxford: Oxford University Press, 1986), 135.

"This, indeed, was a dreadful interval": Burney, "Mastectomy," 136.

"I stood suspended . . .": Burney, "Mastectomy," 137.

They placed an ordinary cambric handkerchief: Burney, "Mastectomy," 138.

"When the dreadful steel . . .": Burney, "Mastectomy," 138–40.

"I must own, to you . . .": Burney, "Mastectomy," 140–41.

Until the advent of chemotherapy: Marios Loukas, et al., "The History of Mastectomy," *American Surgeon* 77.5 (May 2011): 566–71.

surgeon Jean-Louis Petit published: Loukas, et al., "History of Mastectomy."

Both operations removed breasts: Loukas, et al., "History of Mastectomy."

Doctors saw plenty of late-stage tumors: Aronowitz, *Unnatural History*, 91.

"At its most reductive . . .": Ellen Leopold, *A Darker Ribbon: Breast Cancer, Women, and Their Doctors in the Twentieth Century* (Boston: Beacon Press, 1999) 62–63.

Interestingly, while doctors overwhelmingly: Donald Oken, "What to Tell Cancer Patients: A Study of Medical Attitudes," *Journal of the American Medical Association*, 175, no. 13 (April 1, 1961): 1120–28.

"The fact that her surgeon . . .": Leopold, *Darker Ribbon*, 122–25.

Fanny Rosenow called the New York Times: Jimmie Holland and Sheldon Lewis, *The Human Side of Cancer: Living with Hope, Coping with Uncertainty* (repr., New York: Harper Perennial, 2001) 8–9; Mukherjee, *Emperor of All Maladies*, 26–27.

Public discussion of breast cancer: Women's cancer: Nancy G. Brinker with Joni Rodgers, *Promise Me: How a Sister's Love Launched the Global Movement to End Breast Cancer* (New York: Crown Archetype, 2010) 169, 212, 214; Corinna Wu, "A Leading Lady," *Cancer Today*, American Association for Cancer Research, September 27, 2012, accessed March 14, 2014, www.cancertodaymag.org/Fall2012/Pages/betty-ford-yesterday-and-today.aspx.

Rose Kushner, who went after: Olson, *Bathsheba's Breast*, 172.

Eventually, she testified before NIH: Hon. Constance A. Morella, "Tribute to Rose Kushner," Congressional Record 101st Congress (1989–1990), Library of Congress, January 23, 1990; Leopold, *Darker Ribbon*, 234–35.

After all, it wasn't the surgeon's life: Olson, *Bathsheba's Breast*, 174.

Now, the two-step biopsy procedure: Judith Rosenbaum, "Rose Kushner: 1929–1990," *Jewish Women: A Comprehensive Historical Encyclopedia*, accessed March 14, 2014, http://jwa.org/encyclopedia/article/kushner-rose.

She spoke the unspeakable words: Brinker, *Promise Me*, 214–16; Wu, "A Leading Lady."

"many Americans still considered . . .": "New Attitudes Ushered in by Betty Ford," *New York Times*, October 17, 1987.

Within a few weeks of Ford's diagnosis: Jane E. Brody, "Inquiries Soaring on Breast Cancer," *New York Times*, October 6, 1974, 21.

According to her 2011 Time *magazine obituary:* Nancy Gibbs, "Betty Ford, 1918–2011," *Time*, July 8, 2011.

A Harris Interactive poll: Emily Wax, "To Cancer 'Previvors,' Angelina Jolie's Public Revelation Was a Gift of Support," *Washington Post*, July 11, 2013; Lisa Szabo, "Angelina Jolie's News Prompts Women to Call Doctors," *USA Today*, May 15, 2013; Liz Neporent, "Jolie's Doctor Says Her Story Raises Awareness, Saves Lives," ABC News, October 15, 2013, http://abcnews.go.com/Health/angelina-jolies-doctor-story-raises-awareness-saves-lives/story?id=20567547.

Unfortunately, the idea that women: Over at Forbes, David Kroll was concerned "that women with breast cancer may think they are not being aggressive enough with their current treatment plan," as well as that Jolie's mastectomy might scare some women off from yearly mammograms and self-examinations. (David Kroll, "A Cautionary Perspective on Angelina Jolie's Double Mastectomy," *Forbes*, May 14, 2013. Accessed April 18, 2014. www.forbes.com/sites/davidkroll/2013/05/14/a-cautionary-perspective-on-angelina-jolies-double-mastectomy/.); At CNN.com, H. Gilbert Welch noted, "If American women saw themselves in Angelina Jolie—then that would be a problem. Because the next logical question is: Should I get a preventive mastectomy?" (H. Gilbert Welch, "What Angelina Jolie forgot to mention," *CNN.com*, May 18, 2013. Accessed April 18, 2014. www.cnn.com/2013/05/17/opinion/welch-jolie-mastectomy/.); At Salon, Joan Walsh concluded her piece on Jolie with, "It would be wonderful if Jolie's essay motivated more women to learn more about their breast cancer risk. It would be sad if it scared them out of screening, or into getting unnecessary mastectomies." (Joan Walsh, "Angelina Jolie's Choice Not the Only One,"

Salon, May 14, 2013. Accessed April 18, 2014. www.salon.com/2013/05/14/angelina_jolie%E2%80%99s_choice_not_the_only_one/.) For more examples of how the media treated Jolie's mastectomy, simply do an Internet search for "Jolie unnecessary mastectomy."

Some of the articles confused: It is worth noting that between 1993 and 2003, the rates of women with breast cancer—not necessarily just BRCA patients—choosing contralateral prophylactic mastectomy increased by about 150 percent, a trend that has justly concerned medical professionals. (Abenaa M. Brewster and Patricia Parker, "Current Knowledge on Contralateral Prophylactic Mastectomy Among Women with Sporadic Breast Cancer," *Oncologist*, 16(7), (July 2011): 935–941.) Still, although it appears that contralateral mastectomy may not increase overall life span, I feel that dubbing these mastectomies "unnecessary" does an injustice to the women who choose them. Life span is not the only measure of an operation's effectiveness and worthiness. For women without BRCA mutations who have already had cancer in one breast, and who are under fifty at the time of their first diagnosis, the risk of developing contralateral breast cancer is around 13 percent—about the same as an average woman's chance of developing the disease in the first place. These women have already failed their first dice roll—is it so unreasonable for these same patients to worry about losing the roll again during round two? Who gets to decide what level of lifetime risk—what level of fear of breast cancer—is tolerable for an individual in the long run? Presumably, to these patients, a contralateral mastectomy may feel essential as a tool for remedying fear of a recurrence, as well as for reducing the relatively low risk of developing contralateral cancer.

What science we have shows: Kelly Metcalfe, et al., "Contralateral Mastectomy and Survival after Breast Cancer in Carriers of BRCA1 and BRCA2 Mutations: Retrospective Analysis," *BMJ* (February 11, 2014): 348:g226; L. C. Hartmann, et al., "Efficacy of Bilateral Prophylactic Mastectomy in BRCA1 and BRCA2 Gene Mutation Carriers," *Journal of the National Cancer Institute* 93, no. 21 (2001): 1633–37; T. R. Rebbeck, et al., "Bilateral Prophylactic Mastectomy Reduces Breast Cancer Risk in BRCA1 and BRCA2 Mutation Carriers: The PROSE Study Group," *Journal of Clinical Oncology* 22, no. 6 (2004): 1055–62; T. C. van Sprundel, et al., "Risk Reduction of Contralateral Breast Cancer and Survival after Contralateral Prophylactic Mastectomy in BRCA1 or BRCA2 Mutation Carriers," *British Journal of Cancer* 93, no. 3 (2005): 287–92; D. G. R. Evans, et al., "Risk Reducing Mastectomy: Outcomes in 10 European Centres," *Journal of Medical Genetics* 46, no. 4 (2009): 254–58; S. M. Domchek, et al., "Association of Risk-Reducing Surgery in BRCA1 or BRCA2 Mutation Carriers with Cancer Risk and Mortality," *Journal of the American Medical Association* 304, no. 9 (2010): 967–75; A. Finch, et al., "Salpingo-oophorectomy and the Risk of Ovarian, Fallopian Tube, and Peritoneal Cancers in Women with a BRCA1 or BRCA2 Mutation," *Journal of the American Medical Association* 296, no. 2 (2006): 185–92; S. L. Ingham, et al., "Risk-Reducing Surgery Increases Survival in BRCA1/2 Mutation Carriers Unaffected at Time of Family Referral," *Breast Cancer Research and Treatment* 142, no. 3 (December 2013): 611–18; R. Calderon-Margalit and O. Paltiel, "Prevention of Breast Cancer in Women Who Carry BRCA1 or BRCA2 Mutations: A Critical Review of the Literature," *International Journal of Cancer* 112, no. 3 (November 10, 2004): 357–64.

Rose Kushner and fellow journalist Betty Rollin: Mukherjee, *Emperor of All Maladies*, 200; Olson, *Bathsheba's Breast*, 175–76.

He spent ten years gathering data: Mukherjee, *Emperor of All Maladies*, 200.

Doctors prescribed oophorectomy: Löwy, *Preventive Strikes*, 20.

Irregular period? Epileptic? Nymphomaniac?: Leopold, *Darker Ribbon*, 61.

the removal of ovaries has been linked: D. Shoupe, et al., "Elective Oophorectomy for Benign Gynecological Disorders," *Menopause* 14, no. 3 (May–June 2007, part 2): 580–5.

Removing the uterus through hysterectomy: "Hysterectomy," Medline Plus, National Institutes of Health, accessed March 14, 2014, www.nlm.nih.gov/medlineplus/hysterectomy.html.

An 1882 paper by New York surgeon: T. G. Thomas, "On the Removal of Benign Tumors of the Mamma without Mutilation of the Organ," *N.Y. State J. Med. Obstet. Rev.* (April 1882): 337–40.

Thirty-five years later, St. Louis surgeon Willard Bartlett: Willard Bartlett, "An Anatomic Substitute for the Female Breast," *Annals of Surgery* (Philadelphia) 66 (1917): 208–11.

The idea that cystic mastitis: Mayo Clinic Staff, "Fibrocystic Breasts," Mayo Clinic, March 8, 2013, accessed March 14, 2014, www.mayoclinic.com/health/fibrocystic-breasts/DS01070.

Surgeon Joseph Bloodgood: "Joseph Cold Bloodgood: November 1, 1867–October 22, 1935," *American Journal of Cancer* 26 (1936): 397–98.

Cancer and lumpy breasts feel: J. C. Bloodgood, "Border-line Breast Tumors," *Annals of Surgery* 93 (1931): 239, 249.

From this he concluded: Bloodgood, "Border-line Breast Tumors," 246–47.

"The danger of an incomplete operation . . .": Bloodgood, "Border-line Breast Tumors," 241.

Faced with uncertainty in 1931: Bloodgood, "Border-line Breast Tumors," 235–49.

By the 1940s: Löwy, *Preventive Strikes*, 97.

A mid-1960s survey found: Nora Jacobson, *Cleavage: Technology, Controversy, and the Ironies of the Man-Made Breast* (New Brunswick, NJ:

Rutgers University Press, 2000) 129. That research is based on a survey cited in Michael Gurdin and Gene A. Carlin, "Complications of Breast Implantations," *Plastic and Reconstructive Surgery* 40, no. 6 (1967): 530–33 and Tibor de Cholnoky, "Augmentation Mammaplasty: Survey of Complications in 10,941 Patients by 265 Surgeons," *Plastic and Reconstructive Surgery* 45, no. 6 (1970): 573–77.

Prophylactic mastectomy in the 1970s was marked: Jacobson, *Cleavage*, 129–134.

In the meantime, surgeons performed: Ibid., 129–30.

The X-rays picked up stuff: Löwy, *Preventive Strikes*, 107.

As mammography became standard: Ibid.

"In the 1970s, women diagnosed . . .": Ibid., 108.

those results depended on having: Jacobson, *Cleavage*, 130.

Pennisi's follow-up study: Jacobson, *Cleavage*, 132–134. That research is based on Vincent R. Pennisi and Angelo Capozzi, "The Incidence of Obscure Carcinoma in Subcutaneous Mastectomy," *Plastic and Reconstructive Surgery* 56, no. 1 (1975): 9–11.

It's hard to get good data: Loukas, et al., "History of Mastectomy."

Chapter 8: The Black Cloud

alcohol consumption has been connected: "What Are the Risk Factors for Breast Cancer?," American Cancer Society, January 31, 2014, accessed March 14, 2014, www.cancer.org/cancer/breastcancer/detailedguide/breast-cancer-risk-factors.

Once known as "nun's disease": Olson, *Bathsheba's Breast*, 21.

A professor of medicine writing: Ibid., 22.

Women who give birth: Ibid.

women who have fewer periods during their lifetimes: "Pregnancy and Breast Cancer," American Cancer Society, October 2, 2013, accessed March 14, 2014, www.cancer.org/cancer/bresatcancer/moreinformation/pregnancy-and-breast-cancer.

Pregnancy reduces ovarian: "Reproductive History and Breast Cancer Risk," National Cancer Institute, National Institutes for Health, accessed March 14, 2014, http://m.cancer.gov/topics/factsheets/reproductive-history.

While having kids lowers risk: Ibid.

Tamoxifen reduces the risk: Löwy, *Preventive Strikes*, 185.

A 1998 study: Bernard Fisher, et al., "Tamoxifen for Prevention of Breast Cancer: Report of the National Surgical Adjuvant Breast and Bowel Project P-1 Study," *Journal of the National Cancer Institute* 90, no. 18 (1998): 1371–88.

A subsequent smaller trial of tamoxifen: M. King, et al., "Tamoxifen and Breast Cancer Incidence Among Women with Inherited Mutations in

BRCA1 and BRCA2: National Surgical Adjuvant Breast and Bowel Project (NSABP-P1) Breast Cancer Prevention Trial," *Journal of the American Medical Association* 286, no. 18 (2001): 2251–56.

tamoxifen reduces the risk of estrogen receptor–positive: "Risk-Lowering Drugs," Susan G. Komen, January 21, 2014, accessed March 14, 2014, http://ww5.komen.org/BreastCancer/RiskLoweringDrugs.html.

Of course, taking estrogen blockers: "Hormone Therapy for Breast Cancer," National Cancer Institute, National Institutes for Health, accessed March 14, 2014, www.cancer.gov/cancertopics/factsheet/Therapy/hormone-therapy-breast.

a little hormonal birth control: "Oral Contraceptives Reduce Long-Term Risk of Ovarian Cancer," National Cancer Institute, National Institutes for Health, accessed March 14, 2014, www.cancer.gov/cancertopics/prevention/ovarian/oral-contraceptives.

Chapter 9: Barbie Girls

"If she would have seen Newt Gingrich . . .": Gingrich allegedly asked his first wife for a divorce while she was recovering from cancer treatment. Gingrich has disputed this claim. Jeff Zeleny, "On the Stump, Gingrich Puts Focus on Faith," *New York Times,* February 26, 2011, A1. Helen Kennedy, "Report: Newt Gingrich's Divorce Papers Contradict His Claim Wife Jackie Wanted Split," *New York Daily News*, December 26, 2011.

Barbie creator Ruth Handler: Olson, *Bathsheba's Breast*, 235 (quoting from Handler's autobiography).

Her doctor told her: Stocking detail: Elaine Woo, "Ruth Handler, Inventor of Barbie Doll, Dies at 85," *Los Angeles Times*, April 28, 2002, A1.

Handler stood, confused: Veronica Horwell, "Obituary: Ruth Handler: Creator of the Doll Whose Changing Style Defined Generations of Young Women," *Guardian*, May 2, 2002; Woo, "Ruth Handler, Inventor."

"Every woman knows . . .": Olson, *Bathsheba's Breast*, 235.

So Handler decided to remedy: Woo, "Ruth Handler, Inventor"; Horwell, "Obituary: Ruth Handler."

The surprisingly long history: Surajit Bhattacharya, "Sushruta—Our Proud Heritage," *Indian Journal of Plastic Surgery* 42, no. 2 (July–December 2009): 223–25; V. K. Raju, "Susruta of Ancient India," *Indian Journal of Ophthalmology* 51 (2003): 119–22.

He also had a method: Elizabeth Haiken, *Venus Envy: A History of Cosmetic Surgery* (Baltimore: Johns Hopkins University Press, 1997), 5; Bhattacharya, "Sushruta."

The forehead flap: Haiken, *Venus Envy*, 5.

Gasparo Tagliacozzi, who was familiar: Ibid.; Jacobson, *Cleavage*, 172.

The war yielded an unprecedented: Haiken, *Venus Envy*, 29.

Columbia and Johns Hopkins joined in: Ibid., 30.
"miracle man of the Western front": Ibid.
Harold Delf Gillies, one of the fathers: Ibid., 31; Michael Mosley, "How Do You Fix a Face That's Been Blown Off by Shrapnel?," BBC, accessed March 14, 2014, www.bbc.co.uk/guides/zxw42hv.
During the war and after: Haiken, *Venus Envy*, 34.
By the beginning of World War II: Ibid., 34–35.
A month later in Atlantic City: Ibid., 17–18.
Between the Middle Ages and the Renaissance: Yalom, *History of the Breast*, 54.
Upper-class women could afford wet nurses: Ibid., 69–70.
As a 1988 Wall Street Journal article: Ibid., 180–81 (citing a December 2, 1988, piece).
By the 1950s: Jacobson, *Cleavage*, 108.
The "breast fetish" and breast-centric fashions: Yalom, *History of the Breast*, 136,137–138. The nightmare line is a paraphrase of Yalom.
The fad for big breasts: Haiken, *Venus Envy*, 237.
Psychiatrists described typical enhancement: Jacobson, *Cleavage*, 113.
Flat-chested women suffered: Haiken, *Venus Envy*, 271.
"the idea that surgery . . .": Jacobson, *Cleavage*, 116.
Dramatic physical operations could cure: Haiken, *Venus Envy*, 135.
In 1895, Heidelberg surgeon Vincenz Czerny: Jacobson, *Cleavage*, 48; Olson, *Bathsheba's Breast*, 115.
Louis Ombrédanne described a technique: Jacobson, *Cleavage*, 57–58.
Czerny's experimentation with fat: Haiken, *Venus Envy*, 236.
By the 1940s, transplants: Ibid., 236.
Surgeons in the early to mid-1900s: Jacobson, *Cleavage*, 56; S. Bondurant, V. Ernster, and R. Herdman, eds., *Safety of Silicone Breast Implants* (Washington, DC: National Academies Press, 1999), 21, www.nap.edu/openbook.php?record_id=9602&page=21.
Paraffin comes in two varieties: Jacobson, *Cleavage*, 52.
Gersuny diluted his melted: Ibid., 52–53.
By the turn of the century, doctors injected: Ibid., 35; Haiken, *Venus Envy*, 21.
This turned out to be a terrible idea: Jacobson, *Cleavage*, 54–55.
Paraffin wasn't the only injectable: Anne S. Kasper and Susan J. Ferguson, *Breast Cancer: Society Shapes an Epidemic* (New York: St. Martin's Press, 2000) 58–59.
As the story goes: Jacobson, *Cleavage*, 80–81; Haiken, *Venus Envy*, 246.
In 1946 American doctor Harvey D. Kagan: Kasper and Ferguson, *Breast Cancer*, 59; Haiken, *Venus Envy*, 247.
According to a report by the National Academy of Sciences: Bondurant et al., *Safety of Silicone*, 22.
Famed San Francisco topless dancer Carol Doda: Haiken, *Venus Envy*, 247–248.

However, these studies looked at low-volume: Jacobson, *Cleavage*, 81; Haiken, *Venus Envy*, 247.

It also got into the lymphatic system: Haiken, *Venus Envy*, 249.

If accidentally injected into the blood: Kasper and Ferguson, *Breast Cancer*, 60.

The FDA had classified silicone: Everett R. Holles, "Silicone Infections Traced to Tijuana," *New York Times*, July 15, 1974, 30; Walter Oeters and Victor Fornasier, "Complications from Injectable Materials Used for Breast Augmentation," *Canadian Journal of Plastic Surgery* 17, no. 3 (Fall 2009): 89–96.

silicone butt injections: Lizzie Crocker and Caitlin Dickson, "Illegal Butt Injections Are on the Rise and Women Are at Risk," *Daily Beast*, October 13, 2012; Laura J. Nelson, "As Demand for Illegal Silicone Injections Grows, So Do Deaths," *Los Angeles Times*, October 9, 2012; Willard L. Cooper, "Buttloads of Pain: Illegal Ass Enhancements May Be America's Next Health Epidemic," *Vice*, January 7, 2014, accessed April 16, 2014, www.vice.com/read/buttloads-of-pain-0000190-v21n1.

Although surgeons tried the war-produced materials: Jacobson, *Cleavage*, 62.

A study published a year later: Ibid., 62–63.

Women had problems with drainage: Ibid., 64.

Surgical resident Frank J. Gerow: John A. Byrne, *Informed Consent: A Story of Personal Tragedy and Corporate Betrayal . . . Inside the Silicone Breast Implant Crisis* (McGraw Hill, 1997), 45–48.

Thirty years later: Kasper and Ferguson, *Breast Cancer*, 61.

Dow Corning launched the Silastic: Leopold, *Darker Ribbon*, 264.

The Silastic model featured: Haiken, *Venus Envy*, 256; Byrne, *Informed Consent*, 48.

Soon, implants of different shapes: Haiken, *Venus Envy*, 256.

In 1976, Congress passed the Medical Device Amendments: Jacobson, *Cleavage*, 16–19; "Classification of Devices Intended for Human Use," *United States Code, 2010 edition*, 12 U.S.C., Title 21, Chapter 9, Part A, Section 360c (Washington, DC: US Government Printing Office), www.gpo.gov/fdsys/pkg/USCODE-2010-title21/html/USCODE-2010-title21-chap9.htm.

Even more alarming, case reports: Jacobson, *Cleavage*, 99–101, 22–24.

Medical societies suspected: Ibid., 25–33.

Augmentation patients would comprise: Ibid., 46–47; Haiken, *Venus Envy*, 230–231.

the panel's decision left out: Haiken, *Venus Envy*, 231.

In 1998, Dow settled: Yalom, *History of the Breast*, 237; "Chronology of Silicone Breast Implants," *Frontline*: Breast Implants on Trial, February 27, 1996, accessed March 14, 2014, www.pbs.org/wgbh/pages/frontline/implants/cron.html.

the FDA cautions: However, in January 2011 the FDA announced that there is a "possible association" between breast implants and anaplastic large-cell lymphoma, which isn't breast cancer, but rather is cancer of the scar capsule that encloses implants. Only about sixty women worldwide have developed this cancer, out of five to ten million women with implants—less than 0.001%. For more on the conclusions that breast implants are safe and do not cause any disease, despite their common secondary complications such as capsular contracture, rupture, and replacement, see: S. Bondurant, et al., *Safety of Silicone Breast Implants*; "Risks of Breast Implants," U.S. Food and Drug Administration, September 25, 2013, accessed March 14, 2014, www.fda.gov/MedicalDevices/ProductsandMedicalProcedures/ImplantsandProsthetics/BreastImplants/ucm064106.htm.

Fourteen years after the original ban: "Regulatory History of Breast Implants in the U.S.," U.S. Food and Drug Administration, September 25, 2013, accessed March 14, 2014, www.fda.gov/MedicalDevices/ProductsandMedicalProcedures/ImplantsandProsthetics/BreastImplants/ucm064461.htm.

"there exists a standard type of breast . . .": Jack Penn, "A Perspective and Technique for Breast Reduction," in Robert M. Goldwyn, ed., *Plastic and Reconstructive Surgery of the Breast* (Boston: Little Brown, 1976), 184 (quoted from Jacobsen, *Cleavage*, 94).

One surgeon in the textbook: Jacobson, *Cleavage*, 94 (quoting Simon Fredericks, "Management of Mammary Hypoplasia," *Plastic and Reconstructive Surgery of the Breast*, 395).

Frank Gerow, half of the team: Florence Williams, *Breasts: A Natural and Unnatural History* (New York: W. W. Norton and Company, 2012), 71.

Chapter 10: Captain Kirk and Doctor Spock

There are oblong incisions: S. Kinoshita, et al., "Clinical comparison of four types of skin incisions for skin-sparing mastectomy and immediate breast reconstruction," *Surgery Today*, September 17, 2013, accessed July 3, 2014, http://www.ncbi.nlm.nih.gov/pubmed/24043394; Grant W. Carlson, "Technical Advances in Skin Sparing Mastectomy," *International Journal of Surgical Oncology* 2011, accessed July 3, 2014, http://www.hindawi.com/journals/ijso/2011/396901/; P.M. Peyser, et al., "Ultra-conservative skin-sparing 'keyhole' mastectomy and immediate breast and areola reconstruction," *Annals of the Royal College of Surgeons of England* 82, no.4 (July 2000): 227–35, accessed July 3, 2014, http://www.ncbi.nlm.nih.gov/pubmed/10932655; S.L. Spear, M.E. Carter, and K. Schwartz, "Prophylactic mastectomy: indications, options, and reconstructive alternatives," *Plastic and reconstructive surgery* 115, no. 3

(March 2005): 891–909, accessed July 3, 2014, http://www.ncbi.nlm.nih.gov/pubmed/15731693.

Chapter 11: Ta-Ta to Tatas

"I suspect that cancer doesn't give a rat's ass . . .": Molly Ivins, "Who Needs Breasts, Anyway?," *Time*, February 10, 2002.

Emotionally, I was laid out: I have previously written about the "good-bye to boobs" party and the run-up to surgery here: Lizzie Stark, "Goodbye to My Breasts," *Daily Beast*, April 24, 2010, http://www.thedailybeast.com/articles/2010/04/24/good-bye-to-my-breasts.html.

She is from the Reach to Recovery program: "About RRI," Reach to Recovery International, accessed March 14, 2014, www.reachtorecoveryinternational.org/about-rri/.

Chapter 12: Heffalumpless

When a 2010 study comes out: Steven A Narod, Kelly Metcalfe, Shelley Gershman, Parviz Ghadirian, et al. "Contralateral Mastectomy and Survival after Breast Cancer in Carriers of BRCA1 and BRCA2 Mutations: Retrospective Analysis," *BMJ* (February 11, 2014): 348:g226; L. C. Hartmann, T. A. Sellers, D. J. Schaid, et al., "Efficacy of Bilateral Prophylactic Mastectomy in BRCA1 and BRCA2 Gene Mutation Carriers," *Journal of the National Cancer Institute* 93, no. 21 (2001): 1633–37; T. R. Rebbeck, T. Friebel, H. T. Lynch, et al., "Bilateral Prophylactic Mastectomy Reduces Breast Cancer Risk in BRCA1 and BRCA2 Mutation Carriers: The PROSE Study Group," *Journal of Clinical Oncology* 22, no. 6 (2004): 1055–62; T. C. van Sprundel, M. K. Schmid, R. A. E. M. Tollenaar, et al., "Risk Reduction of Contralateral Breast Cancer and Survival after Contralateral Prophylactic Mastectomy in BRCA1 or BRCA2 Mutation Carriers," *British Journal of Cancer* 93, no. 3 (2005): 287–92; D. G. R. Evans, A. D. Baildam, A. Howell, et al., "Risk Reducing Mastectomy: Outcomes in 10 European Centres," *Journal of Medical Genetics* 46, no. 4 (2009): 254–58; S. M. Domchek, T. M. Friebel, T. R. Rebbeck, et al., "Association of Risk-Reducing Surgery in BRCA1 or BRCA2 Mutation Carriers with Cancer Risk and Mortality," *Journal of the American Medical Association* 304, no. 9 (2010): 967–75; A. Finch, M. Beiner, S. A. Narod, et al., "Salpingo-oophorectomy and the Risk of Ovarian, Fallopian Tube, and Peritoneal Cancers in Women with a BRCA1 or BRCA2 Mutation," *Journal of the American Medical Association* 296, no. 2 (2006): 185–92; S. L. Ingham, M. Sperrin, and D. G. Evans, "Risk-Reducing Surgery Increases Survival in BRCA1/2 Mutation Carriers Unaffected at Time of Family Referral," *Breast Cancer Research and Treatment* 142, no. 3 (December 2013): 611–18; R. Calderon-Margalit and O. Paltiel, "Prevention of Breast Cancer in Women Who Carry BRCA1 or BRCA2 Mutations: A Critical Review of the Literature," *International Journal of Cancer* 112, no. 3 (November 10, 2004): 357–64.

As with all the BRCA risk numbers, research: M. C. King, et al., "Breast and Ovarian Cancer Risks Due to Inherited Mutations in BRCA1 and BRCA2," *Science* 302, no. 5645 (October 24, 2003): 643–46; A. Antoniou, P. D. P. Pharoah, et al., "Average Risks of Breast and Ovarian Cancer Associated with BRCA1 or BRCA2 Mutations Detected in Case Series Unselected for Family History: A Combined Analysis of 22 Studies," *American Journal of Human Genetics* 72, no. 5 (May 2003): 1117–30; Sining Chen and Giovanni Parmigiani, "Meta-Analysis of BRCA1 and BRCA2 Penetrance," *Journal of Clinical Oncology* 25 (2007): 1329–33.

typical woman's lifetime risk: "What Are the Key Statistics about Ovarian Cancer?," American Cancer Society, February 6, 2014, accessed March 14, 2014, www.cancer.org/cancer/ovariancancer/detailedguide/ovarian-cancer-key-statistics.

Even though it's only: "Ovarian Cancer Statistics," Centers for Disease Control and Prevention, October 23, 2013, accessed March 14, 2014, www.cdc.gov/cancer/ovarian/statistics/.

The problem is that only about 20 percent: "Can Ovarian Cancer Be Found Early?" American Cancer Society, February 6, 2014, accessed March 14, 2014, www.cancer.org/cancer/ovariancancer/detailedguide/ovarian-cancer-detection.

And the likelihood of a woman living: "SEER Stat Fact Sheets: Ovary Cancer," Surveillance, Epidemiology and End Results Program; "Table 21.9: Cancer of the Ovary (Invasive)," Surveillance, Epidemiology and End Results Program, National Cancer Institute, National Institutes of Health, accessed March 14, 2014, http://seer.cancer.gov/csr/1975_2010/browse_csr.php?sectionSEL=21&pageSEL=sect_21_table.09.html; "Survival Rates for Ovarian Cancer, by Stage," American Cancer Society, February 6, 2014, accessed March 14, 2014, www.cancer.org/cancer/ovariancancer/detailedguide/ovarian-cancer-survival-rates.

The symptoms are a hypochondriac's: Mayo Clinic Staff, "Ovarian Cancer: Symptoms," Mayo Clinic, November 10, 2012, accessed March 14, 2014, www.mayoclinic.com/health/ovarian-cancer/DS00293/DSECTION=symptoms; "Health Guide: Ovarian Cancer," *New York Times,* November 17, 2012, accessed March 14, 2014, http://health.nytimes.com/health/guides/disease/ovarian-cancer/; "Symptoms of Ovarian Cancer," Centers for Disease Control and Prevention, November 29, 2012, accessed March 14, 2014, www.cdc.gov/cancer/ovarian/basic_info/symptoms.htm; "Symptoms of Ovarian Cancer," National Ovarian Cancer Coalition, accessed March 14, 2014, www.ovarian.org/symptoms.php.

To reduce your risk of ovarian cancer: "Ovarian Cancer Prevention," National Cancer Institute, National Institutes of Health, accessed March 14, 2014, www.cancer.gov/cancertopics/pdq/prevention/ovarian/Patient/page3.

The National Comprehensive Cancer Network: "About NCCN," National Comprehensive Cancer Network, accessed March 14, 2014, www.nccn.org/about/default.aspx; Mary Daly, et al., "Genetic/Familial High-Risk Assessment."

neither of those tools: Mary Daly, et al., Genetic/Familial High-Risk Assessment," HBCO-A, 1 of 2.

Of those ten surgeries: "Promising Screening Tool for Early Detection of Ovarian Cancer," MD Anderson Cancer Center, August 15, 2013, accessed March 14, 2014, www.mdanderson.org/newsroom/news-releases/2013/ca-125-screening-tool-for-ovarian-cancer.html.

Oophorectomy removes the body's main sources: Mary Daly, et al., "Genetic/Familial High-Risk Assessment."

Masha Gessen wrote in Slate *while contemplating:* Masha Gessen, "A Medical Quest: In Which I Gain a Greater Understanding of Breasts and Ovaries," *Slate,* June 14, 2004.

Surgical menopause does all the other things: Mayo Clinic Staff, "Menopause: Causes," Mayo Clinic, January 24, 2013, accessed March 14, 2014, www.mayoclinic.com/health/menopause/DS00119/DSECTION=causes.

Hormone replacement therapy: Ovarian removal itself cuts a premenopausal woman's risk of breast cancer by about half, according to the Mayo Clinic ("Oophorectomy," Mayo Clinic, April 5, 2011, accessed March 14, 2014, www.mayoclinic.com/health/breast-cancer/WO00095.), but certain sorts of hormone replacement therapy has been shown to raise some women's risk of developing breast cancer, according to the American Cancer Society ("What Are the Risk Factors for Breast Cancer?" American Cancer Society). According to the BRCA-advocacy organization FORCE, "because much of the research has been conducted on women who experienced natural menopause, the applicability to women experiencing early surgical menopause is uncertain." ("Surgical Menopause," FORCE, accessed March 14, 2014, www.facingourrisk.org/info_research/risk-management/surgical-menopause/index.php?PHPSESSID=abd77629cea63d915a379fdbceeb77aa).

Some studies have found links: "Press Release: Early Surgical Menopause Linked to Declines in Memory and Thinking Skills," *Science Daily,* January 14, 2013, accessed March 14, 2014, www.sciencedaily.com/releases/2013/01/130114161319.htm; L. T. Shuster, et al., "Prophylactic Oophorectomy in Pre-menopausal Women and Long-Term Health—A Review," *Menopause International* 14, no. 3 (2008): 111–16; F. Atsma, et al., "Postmenopausal Status and Early Menopause as Independent Risk Factors for Cardiovascular Disease: A Meta-analysis,"

Menopause 13, no. 2 (March–April 2006): 265–79; Cathleen M. Rivera, et al., "Increased Cardiovascular Mortality after Early Bilateral Oophorectomy," *Menopause* 16, no. 1 (2009):15–23; W. A. Rocca, et al., "Survival Patterns after Oophorectomy in Premenopausal Women: A Population-Based Cohort Study," *Lancet Oncology* 7, no. 10 (October 2006): 821–28. But in BRCA1 and 2 patients, there is some indication that prophylactic oophorectomy may lengthen life, see: K. Armstrong, et al., "Hormone Replacement Therapy and Life Expectancy after Prophylactic Oophorectomy in Women with BRCA1/2 Mutations: A Decision Analysis," *Journal of Clinical Oncology* 22, no. 6 (March 15, 2004).

at least one study suggests that BRCA positive women: Roxana Moslehi, et al., "Impact of BRCA Mutations on Female Fertility and Offspring Sex," *American Journal of Human Biology* 22, no. 2 (March–April 2010): 201–5.

If I want to have non-BRCA biological children: Genetic/Familial High-Risk Assessment," MS-27.

The first involves in vitro fertilization: "In Vitro Fertilization (IVF)," MedlinePlus, National Institutes of Health, February 26, 2012, accessed March 14, 2014, www.nlm.nih.gov/medlineplus/ency/article/007279.htm; "Pre-implantation Genetic Diagnosis," Genesis Genetics, accessed March 14, 2014, http://genesisgenetics.org/pgd/.

There may or may not be a slight increased risk: Mayo Clinic Staff, "In Vitro Fertilization (IVF)," Mayo Clinic, June 27, 2013, accessed March 14, 2014, www.mayoclinic.com/health/in-vitro-fertilization/MY01648/DSECTION=risks.

Prenatal genetic diagnosis: "Psychosocial Issues in Inherited Breast Cancer Syndromes," National Cancer Institute, National Institutes of Health, accessed March 14, 2014, www.cancer.gov/cancertopics/pdq/genetics/breast-and-ovarian/HealthProfessional/page4#Section_975; Claire Julian-Reynier, et al., "BRCA1/2 Carriers: Their Childbearing Plans and Theoretical Intentions about Having Preimplantation Genetic Diagnosis and Prenatal Diagnosis," *Genetics in Medicine* 14, no. 5 (January 12, 2012): 527–34.

If I'm able, I'd like to have: Kutluk Oktay, et al., "Association of BRCA1 Mutations With Occult Primary Ovarian Insufficiency: A Possible Explanation for the Link Between Infertility and Breast/Ovarian Cancer Risks," *Journal of Clinical Oncology* 28, no. 2 (January 10, 2010): 240–44. On the other hand, here's an epidemiological study that suggests there's no difference in parity or fertility for BRCA women: T. Pal, et al., "Fertility in Women with BRCA Mutations: A Case-Control Study," *Fertility and Sterility* 93, no. 6 (April 2010): 1805–8.

Chapter 13: Through the Looking Glass

she posts a new blog entry: Cheri, *Stella Blue: My Life with Metastatic Breast Cancer* (blog), accessed March 14, 2014, http://stellablue.net/.

fine-needle biopsies do have: "Fine Needle Aspiration (Fine Needle Biopsy)," Susan G. Komen, October 22, 2013, accessed March 14, 2014, http://ww5.komen.org/BreastCancer/FineNeedleBiopsy.html.

A 2004 look at the National Cancer Institute's grants: Clifton Leaf, "Why We're Losing the War on Cancer," *Fortune*, March 22, 2004; Jonathan Sleeman and Patricia S. Steeg, "Cancer Metastasis as a Therapeutic Target," *European Journal of Cancer* 46, no. 7 (May 2012): 1177–80.

up to 30 percent of patients: Elia Ben-Ari, "A Conversation with Dr. Patricia Steeg . . ."; National Cancer Institute, "Ten Years of Tamoxifen . . ."; A.M. Gonzales-Angulo, et al., "Overview of Resistance to Systemic Therapy . . ."; V. Guarneri and P. F. Conte, "The Curability of Breast Cancer and the Treatment of Advanced Disease."

about 5 percent of all breast cancer: National Cancer Institute, "Stage Distribution (SEER Summary Stage 2000), By Cancer Site," Surveillance, Epidemiology, and End Results Program, National Institutes of Health, accessed March 14, 2014, http://seer.cancer.gov/faststats/selections.php?run=runit&output=2&data=1&statistic=12&year=201303&race=1&sex=3&age=1&series=cancer&cancer=553.

A full 16.5 percent of African American: "High-Penetrance Breast and/or Ovarian Cancer Susceptibility Genes," National Cancer Institute, National Institutes of Health, February 20, 2014, accessed March 14, 2014, www.cancer.gov/cancertopics/pdq/genetics/breast-and-ovarian/HealthProfessional/page2.

Index